Sensors and Transducers

Other Macmillan books of related interest

J. C. Cluley, *Transducers for Microprocessor Systems*

B. A. Gregory, *An Introduction to Electrical Instrumentation and Measurement Systems*

M. J. Usher, *Information Theory for Information Technologists*

Sensors and Transducers

M.J. USHER

Department of Cybernetics
University of Reading

MACMILLAN

First published 1985

Published by
Higher and Further Education Division
MACMILLAN PUBLISHERS LTD
Houndmills, Basingstoke, Hampshire RG21 2XS
and London
Companies and representatives
throughout the world

Printed in Hong Kong

British Library Cataloguing in Publication Data
Usher, M. J.
    Sensors and transducers.
    1. Detectors 2. Transducers
    I. Title
    620'.0044 TK7872.D/

ISBN 0-333-38709-0
      0-333-38710-4 Pbk

Contents

CONTENTS

Preface

Most quantities that we need to measure are inherently analogue. There is nothing very digital about a length or a temperature and, even though light may be considered to consist of photons, most measurements involve such large numbers that the process is effectively analogue. Our own senses are analogue, so it is hardly surprising that the vast majority of physical sensors are analogue as well. It is only since the developments in microprocessor technology that digital transducers have become important; however, they still have to measure analogue quantities, and most digital transducers therefore employ exactly the same physical principles as their analogue counterparts.

The words 'sensor' and 'transducer' are widely used in referring to sensing devices, the former having gained in popularity over the latter in recent years. This is a pity because 'transducer' stresses the change of form of energy basic to the sensing process and leads to an elegant and powerful classification of devices. The word 'transducer' is used here when considering a complete sensing device, in which a change of form of energy *always* occurs, the word 'sensor' being reserved for devices that are not energy-converting, such as a thermistor which simply changes its resistance in response to temperature.

The aim of the book is to provide an integrated account of the principles and properties of the most important types of physical transducer. The first chapter discusses the types of physical energy and the corresponding signals, and identifies three basic types of transducer. A synthesis of the subject is attempted in chapter 2, by describing the analogies that exist between different physical systems. Chapter 3 starts with a three-dimensional representation of all possible transducers and goes on to consider the basic physical mechanisms available for transduction. Chapter 4 develops the relevant expressions for amplifiers and transducer bridges that are required before the detailed descriptions of the basic transducers for length, temperature and radiation can be given in chapters 5, 6 and 7. For each of these quantities the physical background and standards are explained, followed by both a theoretical treatment of the basic transducers and a description of their practical design and application. Chapter 8 includes a description of the application of the basic transducers to several fields of measurement such as acceleration, force, pressure and flow, and the final chapter reviews some of the most recent developments in the transducer field, such as optical-fibre sensors, resonator sensors, solid-state devices, and smart sensors.

The book is specifically about transducers rather than measurement systems, and provides an integrated and comprehensive coverage of presently available transducers and recent developments. The classification of transducers, based on energy conversion, automatically encompasses both 'analogue' and 'digital' devices, which are seen to be simply different ways of using a particular physical principle. The book is intended as an undergraduate text for first year students in engineering, physics and information technology.

Acknowledgements

The author wishes to thank Professor Fellgett, Head of the Department of Cybernetics at Reading University, for his suggestions and encouragement regarding the lecture courses on which the book is based, and his wife and Mrs Jean Wild, who did the typing (and corrections). He also acknowledges the assistance of the many Cybernetics students who acted as willing guinea-pigs* during the development of the lecture course.

*Guinea-pigs are nocturnal animals that mostly sleep during the day.

1

Introduction

1.1 Analogue and digital quantities

Recent developments in technology and the availability of cheap microprocessors have led to an increased interest in sensing devices, particularly so-called digital devices suitable for direct interfacing to computer systems. Unfortunately (or perhaps fortunately), the only thing at all digital about human beings is that most of us have ten fingers. We are analogue animals living in an analogue world. The quantities that we need to measure are inherently analogue; they can, in principle, take a continuous range of values, though we may prefer to round the values to whole numbers at some stage. There is nothing very digital about a length or a temperature, and although matter is discrete it is certainly not so to our senses and not so to the vast majority of our sensors. In fact it is quite difficult to think of anything in nature that is inherently digital; almost the only example in measurement is in counting numbers of particles (photons, γ-rays etc.).

Although we may wish to use 'digital' sensors in our computer systems, it is important to realise that they are bound to employ exactly the same physical principles as their analogue counterparts. One sometimes reads of industry lamenting the lack of 'absolute digital sensors' and of the importance of effort directed towards their development. In fact, if one examines a typical 'digital sensor', one usually finds it to be a totally analogue device operated in such a way as to produce a digital output. For example, an optical encoder for angle measurement consists of a disc with opaque and transparent sections, as in figure 1.1(a), producing an output as in figure 1.1(b). The process is perfectly analogue but the output is simply digitised into two levels by a comparator circuit. Similarly a 'digital' sensing device producing a frequency or pulse train proportional to the quantity being sensed is misleadingly named; the measurement becomes digital only when the waveform produced is converted into digits by, for example, a counting circuit. In principle this is exactly similar to digitising an 'analogue'

1

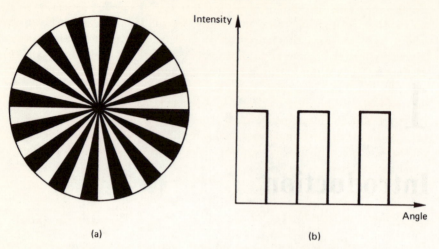

Figure 1.1   *Incremental optical encoder and output signal*

waveform (whose voltage is proportional to the quantity sensed) by an analogue-to-digital converter.

A search for 'absolute digital sensors' is therefore misdirected in principle. What is important is not whether we are using a particular physical property to produce a digital or analogue device, but what the property is and how we can best use it. We need to search instead for suitable physical principles — analogue ones!

This book will concentrate on understanding the basic physical principles available for use in sensing devices. We will see that by a sensible approach to the matter the classifications 'analogue' or 'digital' become hardly necessary, being simply a matter of method of operation, and that it is the underlying physical principles that are all-important.

1.2   Classification of sensing devices

In discussing sensing devices, one has to decide whether to classify them according to the physical property that they use (for example, piezo-electric, photovoltaic etc.) or according to the function that they perform (for example, measurement of length, temperature etc.). In the former case one can present a reasonably integrated view of the sensing process, but it is a little disconcerting when one wishes to compare the merits of, say, two types of temperature sensor, if one has to look through separate sections on resistive, thermoelectric and semiconductor devices to make the comparison. Alternatively, books classifying devices by function often tend to be a rather boring catalogue of numerous unrelated devices. We will try to make a compromise here by first presenting an integrated view of the sensing process in terms of the way in which signals are transformed from one form to another. We will then further synthesise the subject by considering the analogies that exist between the different types of physical system, the various physical properties available for use in sensors and the relations between them. Finally, we will be in a position to discuss sensing devices from the functional viewpoint, under headings such as length, temperature etc., the prior discussion hopefully avoiding the appearance of an interminable list, yet being suitable for someone who actually wants to select or use a sensor for a particular application rather than just read around the subject. The emphasis will be very much on the physical principles of sensing devices, as opposed to complete measurement systems.

1.3 Sensors and transducers

The words 'sensor' and 'transducer' are both widely used in the description of measurement systems. The former is popular in the USA whereas the latter has been used in Europe for many years. The choice of words in science is rather important. In recent years there has been a tendency to coin new words or to misuse (or mis-spell) existing words, and this can lead to considerable ambiguity and misunderstanding and tends to diminish the preciseness of the language. The matter has been very apparent in the computer and microprocessor areas, where preciseness is particularly important, and can seriously confuse persons entering the subject.

The word 'sensor' is derived from *sentire* meaning 'to perceive' and 'transducer' is from *transducere* meaning 'to lead across'. A dictionary definition (*Chambers Twentieth Century*) of 'sensor' is 'a device that detects a change in a physical stimulus and turns it into a signal which can be measured or recorded'; a corresponding definition of 'transducer' is 'a device that transfers power from one system to another in the same or in different form'.

A sensible distinction is to use 'sensor' for the sensing element itself and 'transducer' for the sensing element plus any associated circuitry; for example, a thermistor would be a 'sensor' and a thermistor together with a bridge circuit (to convert resistance change into electrical voltage) would be a transducer. All transducers would thus contain a sensor and most (though not all) sensors would also be transducers. We will use this convention here, though of course the distinction is rather small and as soon as one actually uses a sensor (by applying power to it) it becomes a transducer. An interesting and instructive classification of devices can be achieved by considering the various forms of energy transfer, and the word 'transducer' will therefore be used most often in this book.

Figure 1.2 shows the sensing process in terms of energy conversion. The form of the output signal will often be a voltage analogous to the input signal, though sometimes it may be a waveform whose frequency is proportional to the input or a pulse train containing the information in some other form.

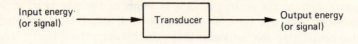

Figure 1.2   *The sensing process*

1.4   Types of transducer

Since the conversion of energy from one form to another is an essential characteristic of the sensing process, it is useful to consider the various forms of energy. The ten forms shown in table 1.1 have been distinguished (Van Dijck, 1964).

Table 1.1   The main forms of energy

Type of energy	Occurrence
Radiant	radio waves, visible light, infra-red etc.
Gravitational	gravitational attraction
Mechanical	motion, displacement, forces etc.
Thermal	kinetic energy of atoms and molecules
Electrical	electric fields, currents etc.
Magnetic	magnetic fields
Molecular	binding energy in molecules
Atomic	forces between nucleus and electrons
Nuclear	binding energy between nuclei
Mass energy	energy given by $E = mc^2$

It is often useful to think in terms of the types of signal associated with the various forms of energy. The essential characteristic of a signal is that of change as a function of time or space, since information cannot be carried or transmitted if a quantity remains constant. Each of the forms of energy has a corresponding signal associated with it, and for measurement purposes six types of signal are important.

(i)   Radiant: especially visible light or infra-red.
(ii)   Mechanical: displacement, velocity, acceleration, force, pressure, flow etc.
(iii) Thermal: temperature, heat flow, conduction etc.
(iv)   Electrical: voltage, current, resistance, dielectric constant etc.
(v)   Magnetic: magnetic flux, field strength etc.
(vi)   Chemical: chemical composition, pH value etc. (this does not appear in table 1.1, but is clearly a distinct form of signal, being derived from several forms of energy).

A general form of measurement system is shown in figure 1.3 (after Middelhoek and Noorlag, 1981b). The signal is fed to an input transducer, which changes the

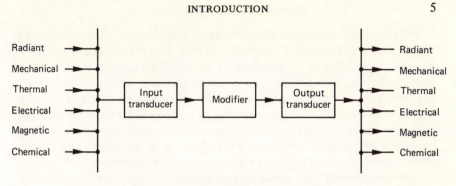

Figure 1.3 *A general measurement system*

form of energy, usually into electrical. The block labelled 'Modifier' represents an amplifier or other device that operates on the transduced signal, and an output ducer then converts the energy into a form suitable for display or recording.

For example, the temperature of a hot body (thermal energy) could be mea-transducer then converts the energy into a form suitable for display or recording. ifier) followed by an LED display (output transducer) producing radiant energy. Similarly, in a photographic exposure meter radiant energy falls on a photoconductive cell, producing a change in resistance; this change is converted to a voltage, by a bridge circuit and amplifier, and produces a deflection of a meter. The overall transduction is between radiant and mechanical energy. A rather different action occurs in a bellows-type pressure gauge; the ambient pressure in the form of mechanical energy produces a deflection of the bellows which may be detected by a lever system, the transduction then being between two forms of mechanical energy (pressure and displacement). One could argue that transduction implies a transformation between forms of energy (as in table 1.1) though the distinction is small here and it is convenient to ignore it.

These three examples illustrate several important points. As mentioned above, we usually want our input transducer to produce an electrical signal that can conveniently be amplified or processed in some way, so the signals on either side of the modifier are usually electrical, the electrical–electrical path being rather trivial. Also, some transducers are said to be self-exciting or self-generating, in the sense that their operation does not require the application of external energy, whereas others, known as 'modulating transducers' do require such a source of energy. A thermocouple is self-generating, producing an e.m.f. in response to temperature difference, whereas a photoconductive cell is modulating. Without an external source of energy a photoconductive cell (effectively a light-dependent resistor) simply responds to the input light energy but does not produce a usable signal; a signal is obtained by applying an electrical voltage and monitoring the resulting current. A third type of transducer is known as a modifier, and is characterised by the same form of energy at both input and output, as in the pressure gauge above. The same word is used in figure 1.3 because the energy form on each side of the modifier is electrical.

Self-generating transducers (thermocouples, piezo-electric, photovoltaic) usually produce very low output energy, having a low effective conversion efficiency; they are often followed by amplifiers to increase the energy level to a suitable value (to drive a meter, for example). In contrast, in modulating transducers, such as photoconductive cells, thermistors or resistive displacement devices, a relatively large flow of electrical energy is controlled by a much smaller input signal energy. Modifying transducers, such as elastic beams or diaphragms, may have a very high conversion efficiency between input and output energy, but usually require some other form of transducer to produce the required electrical output.

The three types of basic transducers are illustrated in figure 1.4.

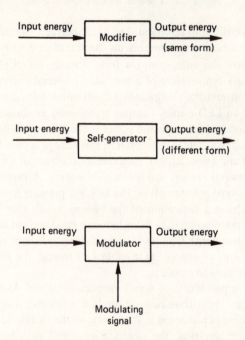

Figure 1.4 *The three types of transducer*

1.5 Transducer parameters

The two most important parameters of a transducer are the output signal produced in response to a given input signal, and the output noise level. The word 'sensitivity' has often been used in connection with transducers, but unfortunately is rather ambiguous. For example, an instrument may be said to have high 'sensitivity' if it produces a large output in response to a given input, or because it is disturbed by people jumping up and down near it, or because it can detect very small input signals. These properties are quite distinct and require separate words to describe them.

The response of a transducer to an input signal is known as its responsivity r, defined by

$$r = \frac{\text{output signal in response to input}}{\text{input signal}}$$

For a displacement transducer the units would be V/m. The definition may be applied between any chosen terminals of the device, so r may be $\Omega/°C$ for a thermistor or $V/°C$ at the output of a bridge circuit connected to the thermistor.

The noise produced by a transducer, which limits its ability to detect a given signal, is known as its detectivity d, defined by

$$d = \frac{\text{signal-to-noise ratio at output}}{\text{input signal}} = \frac{\text{responsivity}}{\text{output noise}}$$

The least detectable signal is defined as that input signal which produces an output r.m.s signal-to-noise ratio of unity (in the chosen output bandwidth), so $d = 1/$ (least detectable signal). The meaning of this is that if, in the absence of an input signal, the transducer produces a certain output noise power N, then the least input signal that can be detected is that which produces an output signal power S equal to N (that is, doubles the total output power). The term 'noise equivalent input signal' is sometimes used instead of 'least detectable signal'. In practice one finds that a signal can be detected with reasonable reliability if the output signal-to-noise ratio is unity, though the 'least detectable signal' is strictly a definition. The units of d are in reciprocal input signal units; for example, 10^6 m^{-1} for a displacement transducer which has a least detectable signal level of 10^{-6} m. It is important to notice that d automatically involves the bandwidth used in the system. Many transducers produce white noise, which has the same power at all frequencies, so the output noise power is proportional to the bandwidth used and d is inversely proportional to the square root of the bandwidth. There are many other parameters such as response time, linearity, accuracy, repeatability, zero-offset, drift, sensitivity to disturbance etc., but they are mostly more applicable to complete measurement systems as opposed to transducers and will not be considered here.

1.6 Exercises on chapter 1

1.6.1. Explain the terms *responsivity, detectivity* and *range*, and give an example of each term by reference to a transducer of your choice for (a) displacement, (b) temperature and (c) radiation (visable or infra-red).

1.6.2 (a) State whether the following transducers are self-generators, modulators or modifiers.

      (i) A rotary potentiometer for angle measurement

      (ii) A thermocouple

      (iii) A photoconductive cell

      (iv) A mercury-in-glass thermometer.

  (b) Give an example of each of the following (do not select those in part (a) above.

      (i) An electrical–thermal–electrical modulator

      (ii) A mechanical–electrical self-generator

      (iii) A mechanical modifier

      (iv) A radiant self-generator

      (v) An electrical–mechanical–electrical modulator.

2

Analogies between Systems

2.1 Analogies

We saw in chapter 1 that transducers operate by transforming energy from one domain to another, such as mechanical to electrical in the case of a piezo-electric device. Interesting analogies exist between several of the basic types of energy or signal, and we will discuss them now in order to illuminate our later discussion and comparison of the types of transducer. It is well known, for example, that the flow of fluid through a pipe is analogous to that of current through a resistor. Consideration of such analogies is not only interesting and instructive in itself, but can have considerable practical application and can sometimes provide the insight required for the solution of a problem by transposing the problem into a more familiar domain. We will consider initially only mechanical and electrical systems, but later extend our view to all the types of energy and signal considered above. The reader is referred to the book by Shearer *et al.* (1971) for a comprehensive treatment of mechanical and electrical networks.

2.2 Mechanical and electrical systems

Figure 2.1 shows a simple mechanical system, comprising a mass M supported by a spring S; in practice some damping is always present and it is usual to indicate this schematically by the dashpot D (even when the only damping is air damping).

We will assume that the mass is constrained to move only vertically by means of frictionless rollers (not shown) and that a force f is applied to it by, say, a magnet attached to it and a coil fixed to the frame (again not shown). The velocity of the mass (and of the spring and dashpot) with respect to the frame is v.

The mass, spring and dashpot are the basic 'building blocks' of any mechanical system and are known as the *elements* of the system. There are only three such

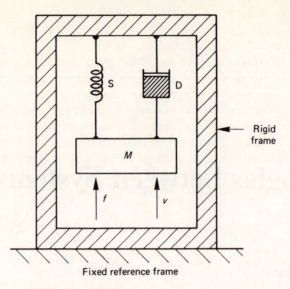

Fixed reference frame

Figure 2.1 *Simple force-driven mechanical system*

passive mechanical elements; note that two of them, the mass and the spring, can store energy but the third, the dashpot, dissipates energy. The force f and the motion v are known as the *variables* of the system (we could choose displacement d or acceleration a, of course).

Each element in a mechanical system is defined by an equation relating it to the variables v and f.

mass M: $f = M \dfrac{dv}{dt}$

spring S: $f = K_m d$ or $\dfrac{df}{dt} = K_m v$ (K_m is known as the stiffness of the spring)

Alternatively

$d = C_m f$ or $v = C_m \dfrac{df}{dt}$ (where $C_m = 1/K_m$ is the compliance of the spring)

dashpot D: $f = R_m v$, assuming viscous damping (R_m is the viscous damping coefficient)

The motion of the mass is given by the differential equation

$$f = M \frac{dv}{dt} + R_m v + \frac{1}{C_m} \int v \, dt$$

Its kinetic energy is $\frac{1}{2}Mv^2$, the potential energy of the spring $\frac{1}{2}C_m f^2$ and the instantaneous power fv.

We will now consider a simple electrical system, developing similar equations with a view to deciding which (if any) of the variables and elements of the two systems can be considered to be analogous.

Figure 2.2 shows such a system, in which an e.m.f. e drives a current i through a series combination of an inductor, capacitor and resistor, producing a voltage $V\ (= e)$ across the network.

Resistors, capacitors and inductors are the three basic passive building blocks in electrical systems, and are clearly the electrical elements; note that again there are two storage elements (capacitance and inductance) and one dissipative element (resistance). Similarly e.m.f. e (or voltage V) and current i (or charge q) are the variables.

The defining equations for the elements are

resistance R : $V = iR$

or $i = GV$ (in terms of conductance $G = 1/R$)

capacitance C: $i = C\dfrac{\mathrm{d}V}{\mathrm{d}t}$

inductance L : $V = L\dfrac{\mathrm{d}i}{\mathrm{d}t}$

The corresponding differential equation for the system is

$$V = L\,\frac{\mathrm{d}i}{\mathrm{d}t} + Ri + \frac{1}{C}\int i\,\mathrm{d}t$$

The energy in the inductor is $\frac{1}{2}Li^2$, in the capacitor $\frac{1}{2}CV^2$ and the instantaneous power is iV.

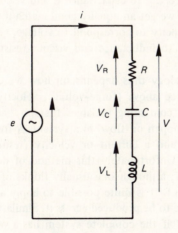

Figure 2.2 *Simple electrical system*

For comparison purposes the equations have been grouped together in table 2.1, choosing the arrangement arbitrarily at present (if you believe that you'll believe anything!).

Table 2.1   Defining equations of mechanical and electrical elements

	Storage		*Dissipative*
	mass M	spring S	dashpot D
Mechanical	$f = M\dfrac{\mathrm{d}v}{\mathrm{d}t}$	$\dfrac{\mathrm{d}f}{\mathrm{d}t} = K_\mathrm{m} v$	$f = R_\mathrm{m} v$
		$v = C_\mathrm{m}\dfrac{\mathrm{d}f}{\mathrm{d}t}$	
	$E = \tfrac{1}{2}Mv^2$	$E = \tfrac{1}{2}C_\mathrm{m}f^2$	$P = fv$
	inductance L	capacitance C	(resistance R) (conductance G)
Electrical	$V = L\dfrac{\mathrm{d}i}{\mathrm{d}t}$	$i = C\dfrac{\mathrm{d}V}{\mathrm{d}t}$	$V = iR$ $i = GV$
	$E = \tfrac{1}{2}Li^2$	$E = \tfrac{1}{2}CV^2$	$P = iV$

Things are usually imperfect in the physical world (often the mental world is not much better) and this topic is no exception; there are apparently two solutions! If we consider that force f corresponds to voltage V and velocity v to current i, we find that the equations are comparable if mass M is analogous to inductance L, compliance C_m to capacitance C and viscous damping R_m to resistance R. Alternatively, we get an equally good match if we take force to correspond to current and velocity to correspond to voltage, giving mass analogous to capacitance, compliance to inductance and viscous resistance to electrical conductance.

The form of the analogy thus depends on how we choose to compare the variables. The first choice above (force–voltage, velocity–current) is known as the force/flow analogy, comparing variables that physically have the effect of 'forcing' something to happen or 'flow' in a system. A force or e.m.f. can clearly be considered 'forcing' and a current or velocity (resulting from a force) are clearly 'flow' variables. Unfortunately this method of deciding on the type of variable is not foolproof; although one usually thinks of current flowing in response to an applied e.m.f. it is quite possible to apply a current generator to a circuit, 'forcing' a voltage to be produced across it. Similarly, a mechanical system may be 'velocity driven' if the complete system has a velocity impressed on it, as in an accelerometer where the frame is moved, producing a force on the suspended mass and a resulting motion.

The second choice above (force–current, velocity–voltage) is known as the through/across analogy, variables being compared in terms of whether they act 'through' or 'across' the system. There is no ambiguity here; strictly an across variable is one that has to be measured between two points in space (for example, voltage or displacement, since a displacement is always with respect to some frame of reference) and a through variable one that can be measured at one point in space (for example, current or force).

The two analogies are summarised in table 2.2. The force/flow analogy is initially more obvious physically, but has the unfortunate consequence that, since it is not based on precise mathematical concepts, when using it one finds the elements that are in series in one domain (for example, mechanical) appear in parallel in the other! As an example the two analogies for figure 2.1, the force-driven mass/spring system, are shown in figure 2.3.

Table 2.2   Mechanical/electrical analogies

Analogy	Variables	Elements
force/flow	force–voltage velocity–current	mass–inductance compliance–capacitance resistance–resistance
through/across	force–current velocity–voltage	mass–capacitance compliance–inductance resistance–conductance

There are a few points to note in drawing these analogies. A fixed reference frame has no motion and is equivalent to electrical ground, and similarly a rigid mechanical beam or connector is equivalent to an electrical conductor of zero resistance. In drawing an analogy for a mass, there is a slight problem since capacitors and inductors have two terminals but a mass has only one (any connection is equivalent to any other). However, the essential point is that the position of a mass is always measured with respect to some reference frame, so one end of the equivalent capacitance or inductance must be connected to electrical ground. (This means that mechanical analogies of some electrical circuits do not exist, as it is quite possible to have an inductor or capacitor with neither terminal grounded.)

In the mechanical system the three elements are clearly in parallel, since they all move with the same velocity (at first sight the mass appears in series, but this is because it has only one terminal — it has an 'implied' terminal connected permanently to a fixed reference frame because its position is always measured with respect to such a frame). The elements appear similarly in parallel in the through/across case (since they all have the same voltage) but in series in the force/flow case (since they all have the same current). In the former there is a simple one-to-one replacement of elements. However the three sets of equations are absolutely identical in form so both analogies are equally valid.

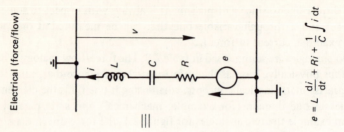

Electrical (force/flow)

$$e = L\frac{di}{dt} + Ri + \frac{1}{C}\int i\,dt$$

$\equiv$

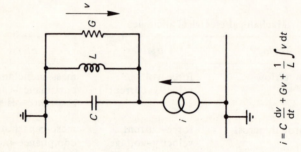
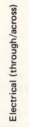

Electrical (through/across)

$$i = C\frac{dv}{dt} + Gv + \frac{1}{L}\int v\,dt$$

$\equiv$

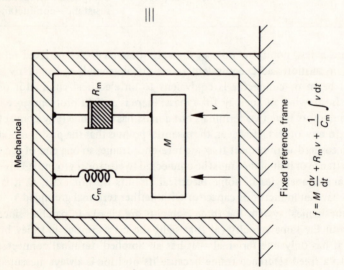

Mechanical

$$f = M\frac{dv}{dt} + R_m v + \frac{1}{C_m}\int v\,dt$$

Fixed reference frame

Figure 2.3 Electrical analogues of mass/spring system

There is a special relationship between the two circuits of figure 2.3, which are redrawn in figure 2.4 to illustrate this.

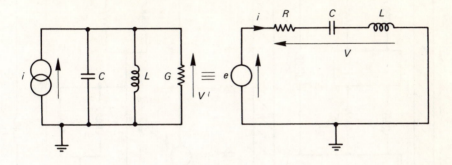

Figure 2.4 *Dual circuits of figure 2.3*

Any electrical circuit has a dual, in which all variables and elements are replaced by their duals, parallel elements appearing in series, meshes becoming nodes etc., but the defining equations always remain identical in form. Our problem of the two analogies is thus resolved — they are simply duals of one another. The duals for electrical and mechanical elements and variables are shown in table 2.3.

Table 2.3   Dual variables and elements

	Variables	
Electrical	$i \leftrightarrow V$	$C \leftrightarrow L$, $R \leftrightarrow G$
Mechanical	$f \leftrightarrow v$	$M \leftrightarrow C_m$, $R_m \leftrightarrow G_m$

Finally, figure 2.5 shows the through/across analogy of a fairly complicated velocity-driven system, illustrating the direct one-to-one correspondence between elements. It would be much harder to use the force/flow method, though the result could be deduced by finding the dual of the electrical circuit. Note that we could easily analyse the electrical circuit using the usual circuit techniques, which would be far easier than attempting an analysis of the original mechanical arrangement. It is, in fact, a type of cascaded low-pass filter.

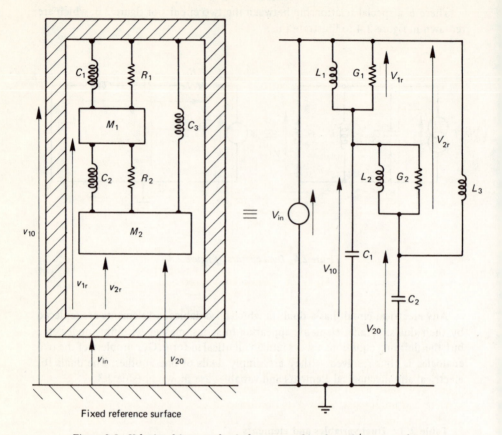

Figure 2.5 *Velocity-driven mechanical system and its through/across analogue*

2.3 Fluid systems

A fluid system is of course mechanical, but we will consider such systems separately since they are rightly a class of their own.

Figure 2.6 shows a simple fluid system comprising a tank of water maintained at a constant height (by a tap and overflow), feeding a second tank via a narrow tube. Water escapes from the second tank by another narrow tube.

The variables in a fluid system are pressure p and flow i (volume/s). The most obvious elements are fluid capacitance C_f and fluid restrictance R_f (or conductance G_f).

The basic equation for restrictance is $p = iR_f$, clearly analogous to Ohm's law, so we have the analogies pressure–voltage, flow–current and restrictance–resistance. In this case the force/flow and through/across classifications give the same result (it is only in the mechanical case above that problems arise). The volume v of a con-

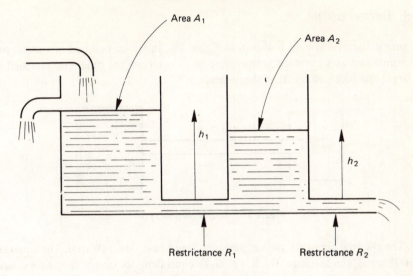

Figure 2.6 *Simple fluid system*

tainer of height h and cross-sectional area A is $v = Ah$, and comparing this with $q = CV$ for electrical charge we have volume–charge, area–capacitance and height–voltage. It is therefore easier to define restrictance by $h = iR_f$ and use height instead of pressure when dealing with simple tanks of fluid. There is an equivalent of electrical inductance, known as fluid inertance, arising because fluid has mass which must be accelerated in moving it. However, in many cases the effect is small and can be neglected The electrical analogue of the system of figure 2.6 (excluding inertance) is shown in figure 2.7. Note that fluid capacitance is represented by an electrical capacitor with one end grounded as before. Dual circuits can be drawn, but the concept is of little use here since inertance is little used.

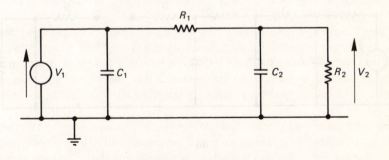

Figure 2.7 *Electrical analogue of fluid system*

2.4 Thermal systems

A simple thermal system is shown in figure 2.8, in which one end of a metal rod is maintained at a constant temperature in an oven and the other end attached to a large solid block at constant temperature.

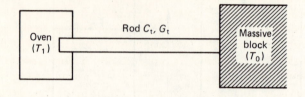

Figure 2.8 *Simple thermal system*

The thermal elements are temperature T and heat flow i (Watts). The equation for thermal conductance G_t is $i = G_t T$, equivalent to Ohm's law, so we have heat flow–current, thermal conductance–conductance and temperature (strictly temperature difference)–voltage. Both variable classifications give the same result. The equation for thermal capacitance C_t is $H = C_t T$, where H is heat energy (as

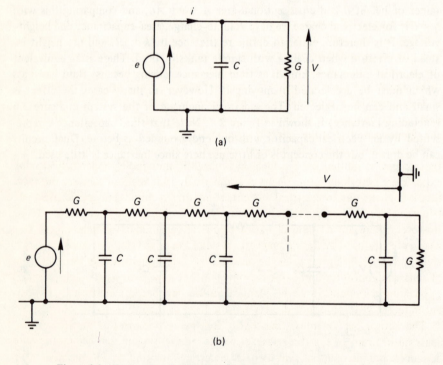

Figure 2.9 *'Lumped' and 'distributed' electrical equivalents of figure 2.8*

opposed to flow); this is directly analogous to $Q = CV$ again, so we have heat–charge and thermal capacitance–capacitance. However, if we look for a property similar to inertia to find a second storage element, rather surprisingly there isn't one! Thermal energy is exhibited by the motion of the electrons in the material and no 'acceleration' is required. This makes thermal equivalents very simple; also, it means that thermal duals do not exist!

An electrical 'equivalent' of figure 2.8 is shown in figure 2.9(a). The reason for calling it an 'equivalent' is that strictly it is not! Thermal capacitance and conductance are 'distributed' parameters; they cannot be represented by a single value at a point, unlike electrical capacitance and conductance. A more accurate 'equivalent' is shown in figure 2.9(b), where the numbers of resistors and capacitors tend to infinity. The equations for the exact equivalent can be found from electrical circuit theory, but the error in using the 'lumped' version of figure 2.9(a) is actually quite small.

If we replaced the oven in figure 2.8 by a heater putting a constant power into the rod, we could get a similar equivalent but driven by a current source instead of a voltage source.

2.5 Other systems: radiant, magnetic, chemical

Systems employing optical radiation are often essentially identical with the thermal systems discussed above. If the radiation is that from a hot body, there will be a stream of radiation emitted from the body and a detector intercepting it will rise in temperature according to its conductance and capacitance, as in the equations above. The same applies to a beam of radiation from, say, a laser, though if it is detected by a photovoltaic or photoconductive detector (that is, a photon detector as opposed to a thermal detector) the analogy is no longer useful since the temperature rise of the detector is unimportant.

Distinct similarities exist between magnetic and electrical quantities, though the analogy is limited by the fact that magnetic monopoles apparently do not exist. They have long been sought experimentally, but without success. Magnetic flux ϕ is analogous to electrical current, being driven round a magnetic circuit of reluctance R_1 by a magnetomotive force M, according to the equation $M = R_1\phi$. However, there are no equivalents of electrical capacitance or inductance. The analogy is useful in analysing magnetic systems, such as velocity transducers. An analogy can be set up between chemical and electrical quantities, in terms of flow of ions and concentration gradients, but does not appear to have much practical application at present. However, chemical sensors have become increasingly important in recent years, and such devices are discussed in chapter 9.

The author has an interesting theory that many non-physical systems can be analysed in terms of the three basic elements representing inertia, storage (or capacitance) and dissipation, and two variables representing force and flow. For example, factory assembly lines and mining involve these terms to different degrees. The concepts apply particularly well to the educational process in students, in which knowledge is the flowing variable and the lecturer the forcing variable. Students clearly display considerable inertia (in getting down to their studies), enormous dissipation (in forgetting taught facts) and almost zero capacity (for learning new ones)!

2.6 Exercises on chapter 2

2.6.1. Draw the through/across electrical analogue of the accelerometer of figure 2.10, and then draw its dual circuit. Label each diagram to show the corresponding variables and elements.

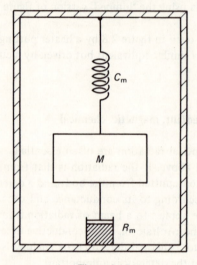

Figure 2.10

2.6.2. Figure 2.11 is a schematic diagram of the suspension system in a car.

    (i) Draw an electrical analogue of the system, stating the correspondence between variables and elements in the analogy that you have chosen. The mass of the tyre and the main spring may be ignored. Discuss the effects of the various elements in damping the motion on the car.

    (ii) Deduce values of the inductors and capacitors in your electrical circuit, assuming a direct one-to-one correspondence. How could the appropriate value of resistance be determined?

2.6.3. Figure 2.12 shows a temperature-control system for maintaining the temperature of the crystal of an oscillator at $50°C$, the surroundings being at $25°C$. The oven is a small hollow copper cylinder of mass 25 g with a

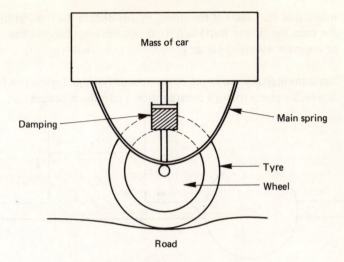

Figure 2.11 *Suspension system of a car (schematic)*

heating coil around it and enclosed by a layer of insulation. The heater operates at a mean power of 2.5 W.

Draw an electrical equivalent of the thermal system (assume that the copper has thermal capacitance only and the insulation thermal resistance

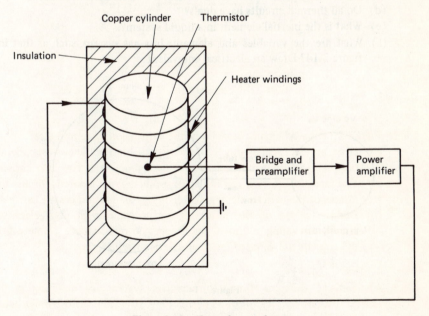

Figure 2.12 *Thermal control system*

only), find the values of the elements, and deduce the transfer function for the oven (in $°C$ per Watt) and its time constant. (Specific heat capacity of copper = 400 J/$°C$ per kg.)

2.6.4. Draw the electrical analogue of the flow-driven liquid system on figure 2.13, in which a pump forces a constant flow i into the container.

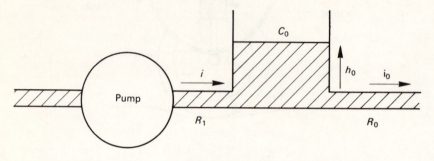

Figure 2.13

What are the effects of restrictances R_1 and R_0?

2.6.5. (a) State a mechanical equivalent of an electrical transformer.
     (b) Can an electrical equivalent be drawn for all mechanical circuits?
     (c) Can a mechanical equivalent be drawn for all electrical circuits?
     (d) Do all thermal circuits have duals?
     (e) What is the inertial element in a liquid system?
     (f) What are the variables and elements in a gas system, such as that in figure 2.14? Draw an electrical equivalent.

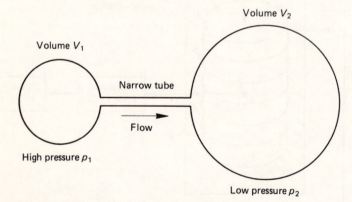

Figure 2.14

3

The Physical Effects available for use in Transducers

3.1 Representation of transducers

An interesting three-dimensional representation of transducers was proposed by Middelhoek and Noorlag (1981b), shown in figure 3.1. The primary energy input to the system is represented by the x-axis and the energy output by the y-axis. Self-generating transducers therefore lie in the x–y plane. With the six forms of energy discussed above we have 36 possibilities, of which 30 are true self-generators and six modifiers (having the same form at both input and output). The most important self-generators are the five having electrical output, shown as crosses in figure 3.1; the six modifiers are indicated by small circles.

The modulating transducers are represented by points in three-dimensional space, the z-component representing the modulating (signal) input. There are evidently 216 modulators in all; the most important are those for which both input and output energy are electrical, of which there are five (shown by dots in the figure), though there are a few known devices with non-electrical x and y components.

There are, of course, a lot of gaps in figure 3.1. However, what is particularly interesting is that the figure apparently can accommodate all known (or possible) transducers, and as new ones are developed the gaps can simply be filled in. It is instructive, though often very difficult or even impossible, to pick a point and try to think of a transducer that could occupy it. The diagram includes both input and output transducers, of course, but we will only be concerned with the former in this book.

Table 3.1 summarises the most important transducers currently in use. However, since chemical energy is very distinct from the other forms (which can all be described as 'physical') and since chemical measurement is a large and important subject in its own right, we have omitted chemical transducers and will not con-

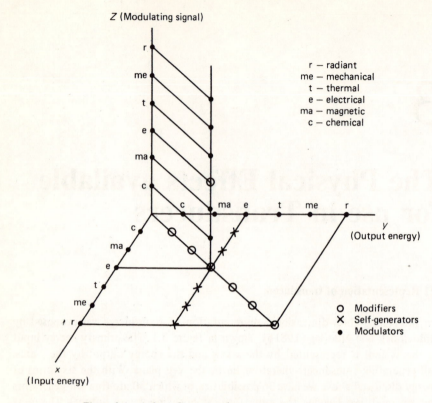

Figure 3.1 *A three-dimensional representation of transducers*

sider them further except for some discussion of recently developed solid-state devices in chapter 9.

We will consider the basic physical processes available for use in transducers in the remainder of this chapter, grouping the processes in terms of transducer type – that is, self-generators, modulators and modifiers. A few applications will be mentioned for completeness, but we will concentrate on the physical principles involved and leave the detailed discussions of specific devices to later chapters.

3.2 Self-generators

3.2.1 Radiant self-generators: the photovoltaic effect

Radiant self-generators are of considerable interest, in addition to their applications in the transducer field, because of the importance of converting radiant energy into electrical energy. Photovoltaic transducers are in fact none other than the well-known silicon solar cells used for satellite power supplies. They are essen-

Table 3.1 The most important physical effects and associated transducers

Type	Transduction	Physical effect	Application
Self-generators	radiant–electrical	photovoltaic; radiation–current	solar cells
	mechanical–electrical	electrodynamic; velocity – voltage	tachogenerators
		piezo-electric; deformation – charge	piezotransducers
	thermal– electrical	thermo-electric; temperature – voltage	thermocouples
		pyro-electric; temperature – charge	radiation detectors
	magnetic–electrical	electromagnetic; flux change – voltage	magnetic field measurements
Modulators	electrical–(radiant) – electrical	photoconductive; radiation – resistance change	radiation detectors
		photo-emissive; radiation – current	
	electrical–(mechanical) – electrical	piezoresistive; strain – resistance change	strain gauges
		displacement – impedance change	electrical displacement transducers
	electrical–(thermal) – electrical	thermoresistive; temperature – resistance change	thermistors, resistance thermometers
	electrical–(magnetic) – electrical	magnetoresistive; magnetic field – resistance change	magnetic field measurement
		Hall effect; e.m.f. due to current in magnetic field	Hall effect probes for current or magnetic field
Modifiers	radiant–(mechanical) – radiant	radiation change due to motion	optical encoders and gratings
	radiant–radiant	temperature change due to collected radiation	thermal radiation detectors
	mechanical–mechanical	displacement change due to pressure	diaphragm pressure transducers
		displacement change due to force	force transducers
		pressure change due to flow	orifice-type flow transducers
	thermal–thermal	temperature change due to heat flow	heat flux detectors
	electrical–electrical	change in electrical form	amplifiers, filters, bridge circuits

tially semiconductor diodes in which light is permitted to fall on the junction region, and are indistinguishable from diodes in the dark (apart from a high reverse current).

When a junction diode is produced, the positive charge carriers in the p material tend to flow to the n material, and similarly for the n material. The p material becomes negatively charged and the n material positively charged, so an electric field (the junction field) is developed in the junction region to stabilise the flow of charge, directed as shown in figure 3.2.

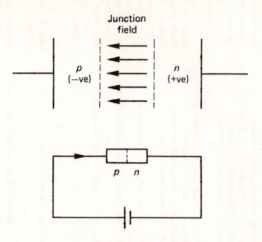

Figure 3.2   p–n *diode*

Forward bias reduces the field and increases current flow, and reverse bias increases the field; the current/voltage characteristic is as shown in figure 3.3(a). For negative bias there is a small reverse leakage current I_0 and for positive bias the current increases exponentially. The relationship is given by

$$I = I_0 \left(\exp\!\left(\frac{eV}{kT}\right) - 1 \right)$$

When light falls on a photoconductive material electrons may be excited across the energy gap E_g, provided that the quantum energy $h\nu$ is greater than E_g, so that the conductivity is increased. The electron in the conduction band and the resulting hole in the valance band are no longer tied together and are therefore free to move. In a p–n junction device the free electron–hole pair comes under the influence of the junction field (provided that the photon is absorbed in the junction region) and the hole is swept to the left (to the p material), and the electron to the right. The hole and electron are thus physically separated and may flow in an external circuit. Moreover, the direction of current flow is in the reverse current direction (since the holes go to the left), so the resulting light current I_L is seen as a large increase in the reverse current, as shown in figure 3.3(a).

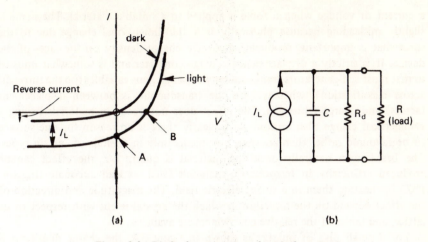

Figure 3.3 *Diode characteristic and equivalent circuit*

The characteristic in the light is simply that in the dark moved downwards bodily by I_L, and this has two important effects. Under short-circuit conditions, at A, the current flow is simply I_L; under open-circuit conditions, at B, a voltage is produced in an external circuit. An equivalent circuit is shown in figure 3.3(b); the light current is shared between the diode resistance R_d and the external load R. There is a significant capacitance C, which may limit the frequency response.

The responsivity is easily evaluated. If we have an incoming stream of mono-chromatic radiation W of frequency v, the number of photons/s is W/hv and the resulting light current is $I_L = qeW/hv$ where q is an efficiency factor. The responsivity r is thus qe/hv A/W, and has a value of about 0.1 A/W at 1μm. Devices are produced by depositing a layer of say p material on a suitable substrate and following this by a thin layer of n material, producing an extended junction. They are usually circular or rectangular and may vary in area between about 1 mm^2 and 10 cm^2. Photovoltaic transducers are particularly valuable, being semiconductor devices and hence directly compatible with modern electronics.

3.2.2 *Mechanical self-generators: the piezo-electric effect*

There are two mechanical self-generators, employing the piezo-electric effect and the electrodynamic effect. However, the latter is closely associated with the electro-magnetic effect and will be considered under magnetic self-generators in section 3.2.4.

The piezo-electric effect

The word 'piezo' means 'push' and the effect is exhibited by the appearance of a current or voltage when a force is applied to a suitable material. The name is slightly misleading because physically it is the dimensional change due to the force that is important, producing a surface charge density on the faces of the device. It is strictly a displacement-to-charge converter and is somewhat unusual in that most self-generators transduce between analogous variables (on the through/ across classification), whereas here the transduction is between displacement (across) and charge (through). The effect does not occur in materials having a symmetrical charge distribution, since clearly there is no reason for one surface to be favoured rather than another, but occurs only in crystals of certain types. The best-known naturally occurring material is quartz but the effect can be produced artificially in ferroelectric materials (such as lead zirconate titanate, PbZ) by heating them in a strong electric field. The magnitude and direction of the effect depend on the direction in which the crystal is cut with respect to its lattice, and tables of the relative coefficients are available.

For a small disc of quartz, as shown in figure 3.4, the charge density q is given in terms of the applied force f by

$$q = df \tag{3.1}$$

where d is known as the 'd coefficient', for want of a better name. d is typically 2×10^{-12} C/N for quarz and about 150×10^{-12} C/N for PbZ. The opposite surfaces of the device are metallised, producing a capacitor of capacitance $C = \epsilon\epsilon_0 A/t$ where ϵ is the relative permittivity (about 4.5 for quartz, 1800 for PbZ), ϵ_0 the permittivity for free space ($10^{-9}/36\pi$ F/m), A the area and t the thickness. With $A = 1$ cm^2 and $t = 1$ mm, we find $C = 4$ pF (1600 pF for PbZ). The voltage corresponding to the charge can be found, and this is often given in terms of the 'g coefficient' in the relationship

$$V = gtP \tag{3.2}$$

where P is the pressure applied. g ($= d/\epsilon\epsilon_0$) has a value of about 5×10^{-2} V m/N for quartz and about 10^{-2} V m/N for PbZ.

Both these basic equations are misleading, causing the device to be thought of as necessarily a force or pressure transducer, whereas a better equation summarising the physical process involved (deformation δt producing charge q) is

$$q = K\delta t \tag{3.3}$$

The constant K can be found in terms of d or g and the bulk modulus E of the material, since $E = (f/A)/(\delta t/t)$

$$K = \frac{q}{\delta t} = \frac{dF}{\delta t} = \frac{dEA}{t} = \frac{dEC}{\epsilon\epsilon_0} = \frac{EC}{g}$$

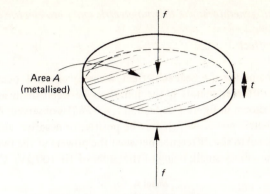

Figure 3.4 *Piezo-electric transducer*

A simple equivalent circuit for a piezo-electric device is shown in figure 3.5(a), choosing a current generator i for convenience (instead of a charge generator) so that i is proportional to velocity $\dot{x}$. The device capacitance C appears in parallel with a large leakage resistance, usually ineffective in practice. The Thévenin equivalent in figure 3.5(b) is also useful, C being placed in series with a voltage generator whose output is proportional to displacement x. The device normally feeds a virtual-earth amplifier, and a response proportional to velocity or displacement (over specific frequency ranges) can be obtained by a suitable choice of bias and feedback.

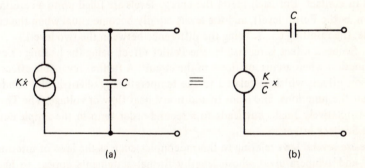

Figure 3.5 *Equivalent circuits of piezo-electric transducer*

Piezo-electric devices are widely used for force and pressure measurement, but it is important to note that in practice their response to displacement does not extend to d.c. because of the series capacitance in figure 3.5(b). They are also widely used for acceleration measurement, a small crystal performing the functions of both supporting spring and detection of relative displacement of the mass. As with most self-generators the effect is reversible, so a displacement results from the application of a charge (or voltage). The piezo-electric effect is usually to blame for the hourly or more frequent noises emitted by digital watches!

3.2.3 Thermal self-generators: the thermo-electric and pyro-electric effects

Thermo-electric effect

This is another name for the Seebeck effect, whereby an e.m.f. occurs in a circuit comprising two different metals if the junctions between them are at different temperatures, as shown in figure 3.6. An e.m.f. $e = P\delta T$ is observed, P being known as the thermo-electric power, which may be positive or negative, and the resulting effect is proportional to the difference between the powers of the two metals used. The magnitudes are fairly small, being of the order of 10–100 μV/°C.

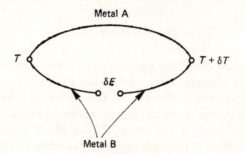

Figure 3.6 *Basic thermocouple*

The effect is due to the equalisation of Fermi levels when two metals are placed in contact. For each metal the energy levels are filled up to a certain value (known as the Fermi level), and the levels rapidly become equal when the contact is made, the resulting e.m.f. being the difference between the two levels.

The Seebeck effect is reversible, the Peltier effect being the heating or cooling of a junction when a current flows in the circuit. A further reversible effect is the Thomson effect, which is related to the temperature gradient in the conductors between the junctions, and leads to additional heat flow or voltage. The Thomson effect is relatively small, but leads to a second-order term in the simple equation for the Seebeck effect.

There are several laws relating to thermocouples, such as the laws of intermediate metals and temperatures, whose lengthy formal statements appear to have delighted some authors in the past, though probably not their readers since these laws are all intuitively obvious.

The pyro-electric effect

This is the thermal equivalent of the piezo-electric effect, in which deformation produces a surface charge density. The word means 'furnace-electricity' and a temperature difference across a disc of pyro-electric material thus produces a charge density. Most piezo-electric materials show the effect, especially semiconductor devices, and it is a serious disadvantage in some cases.

The main practical application is in the measurement of radiation. The stream of radiation is directed on to a small disc of material whose faces are metallised, and a corresponding voltage is produced. The best-known material is lead zirconate **titanate**.

3.2.4   Magnetic self-generators: the electromagnetic and electrodynamic effects

Faraday's law of induction states that the e.m.f. produced in a coil of n turns due to a changing magnetic flux ϕ is

$$e = -n\frac{\mathrm{d}\phi}{\mathrm{d}t}$$

The effect can be used directly for the measurement of changing magnetic fields or of steady fields by rotating the coil at a known rate. However, Faraday's law is most useful for measuring the velocity of a conductor moving in a magnetic field, in which case the transduction action is strictly that of a mechanical self-generator, though it is covered here for completeness. It is often known as the electrodynamic effect.

By considering a straight section of conductor of length l moving with velocity v perpendicular to a magnetic field of induction B, as in figure 3.7, it is easily deduced that an e.m.f. is produced given by

$$e = (Bl)v$$

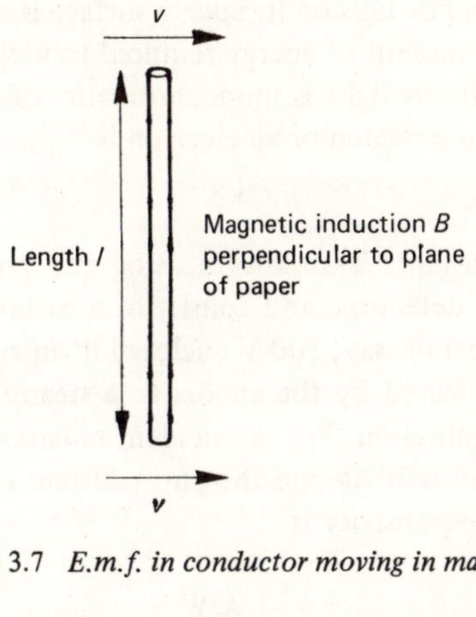

Length l

Magnetic induction B perpendicular to plane of paper

Figure 3.7   E.m.f. in conductor moving in magnetic field

The same formula applies if the conductor is a coil of total length l, and the effect is reversible, so that if a current i is fed to the coil a force $F = (Bl)i$ Newtons is produced. The identity of the coefficients (Bl) in the two cases is very useful in calibrating instruments which use a magnet/coil system for velocity measurement, since a known force can easily be applied by simply adding a small mass.

Most practical devices consist of a fixed magnet with a movable coil attached to the object whose velocity is required, though the reverse configuration is sometimes used. Rotational devices, usually moving coil, are also very widely used, usually being referred to as tachogenerators.

A further application of Faraday's law is in magnetoresistive and Hall-effect transducers, in which the motion of charge carriers in a magnetic field produces a resistance change or e.m.f. These effects are discussed in section 3.3.4 which deals with magnetic modulators.

3.3 Modulators

This is the largest group of transducers. We will consider specifically the five modulators having both electrical input and electrical output, along with a few more general modulators with non-electrical input energy.

3.3.1 Radiant modulators: the photo-electric and photoconductive effects

The photo-electric effect

The photo-electric effect is the emission of electrons from a metal surface when light of a suitable wavelength falls on it. Such a surface is characterised by a work function ϕ, which is the amount of energy required to withdraw an electron from it (that is, to infinity). If the light is monochromatic, of frequency ν and wavelength λ, the condition for emission of an electron is

$$h\nu \geqslant \phi$$

where h is Planck's constant. Transducers using the photo-electric effect are known as photo-emissive detectors, and consist of a cathode of suitable material and an anode at a potential of, say, 100 V enclosed in an evacuated jacket. All the electrons emitted are collected by the anode, so a steady photocurrent flows in response to a steady illumination. For an incident radiation of W watts, the number of photons per second is $W/h\nu$ and the photocurrent is $eqW/h\nu$, where q is an efficiency factor, so the responsivity is

$$r = \frac{eq}{h\nu} \text{ A/W}$$

The responsivity thus increases with wavelength, reaching a maximum for $h\nu = \phi$, after which it rapidly falls to zero. In practice the maximum wavelength is about 1 μm, for a cathode material of caesium oxide and silver.

Photo-emissive detectors are not much used now, but the same effect is employed in photomultipliers, in which the photocurrent is amplified by a series of secondary electrodes, producing very high responsivity and detectivity.

The photoconductive effect

Photoconductive materials are semiconductors in which a transition between valence and conduction bands, separated by an energy gap E_g, may be excited by an incident photon of suitable wavelength. Unlike the photovoltaic effect, in which a physical separation of charge carriers occurs, there is simply a change in conductivity as the name implies. As above, the condition for excitation is $h\nu \geqslant E_g$.

The conductivity of a semiconductor is given by $\sigma = Neu$, where N is the total number of electrons in the conduction band, e the electronic charge and u the mobility of the charge carriers. N is strongly dependent on temperature, according to the formula

$$N = N_0 \exp\left(-E_g/2kT\right) \tag{3.4}$$

where N_0 is the total number of electrons in the material (that is, in the valence and conduction bands) and k is Boltzmann's constant. N therefore increases with temperature and is zero at absolute zero.

If the incident radiant power is W, the number of carriers produced per second is $qW/h\nu$, where q is the efficiency factor. Unlike the photovoltaic effect, the carriers have a limited lifetime τ in the conduction band, and it is easy to show that the steady-state excess carriers δN due to the incident radiation W is equal to $qW\tau/h\nu$. The fractional change in conductivity is the same as the fractional change in bulk resistance, $\delta R/R$, and is

$$\frac{\delta\sigma}{\sigma} = \frac{\delta R}{R} = \frac{qW\tau}{h\nu N} \tag{3.5}$$

It is clear that we require a long lifetime τ to get a large response, though clearly this will limit the frequency response of the device to changing incident light. Also, we require the number of electrons N in the conduction band to be small, so that the element's resistance will be high. The responsivity will increase with λ, as for the photo-emissive device, falling rapidly to zero once the condition $h\nu = E_g$ is reached (that is, for $\lambda > hc/E_g$ where c is the velocity of light). One of the best-known photoconductors is lead sulphide, which responds out to about $3 \ \mu m$.

A further class of photoconductors employs what is known as the charge amplification effect. In some materials, notably cadmium sulphide, a hole-trapping effect occurs owing to impurities (copper ions). The effective lifetime of the carriers is greatly increased and the devices have very high responsivity though very low frequency response.

3.3.2   Mechanical modulators

Strain gauges (and the piezo-resistive effect)

Strain gauges employ the change in resistance of a suitable material when subjected to an applied stress, for the measurement of displacement or strain. The
simplest form of device is a cylinder of area A, diameter d and length l of a
material of resistivity ρ, subject to a force f, as shown in figure 3.8.

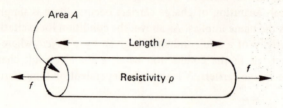

Figure 3.8   *Strain gauge transducer*

The bulk resistance $R = \rho l / A$, so differentiating logarithmically

$$\frac{\delta R}{R} = \frac{\delta l}{l} - \frac{\delta A}{A} + \frac{\delta \rho}{\rho}$$

For a cylinder $A = \pi d^2 / 4$, so $\delta A / A = 2\delta d / d$. The relationship between the change
in diameter and change in length is given by Poisson's ratio $\nu = -(\delta d / d) / \delta l / l$: for
a homogeneous material $\nu = 0.5$. Thus

$$\frac{\delta R}{R} = \frac{\delta l}{l} + \frac{2\nu \delta l}{l} + \frac{\delta \rho}{\rho}$$

The gauge factor (GF) is defined as the ratio of fractional change in resistance to
fractional change in strain

$$\text{GF} = \frac{\delta R / R}{\delta l / l} = 1 + 2\nu + \frac{\delta \rho / \rho}{\delta l / l} \tag{3.6}$$

The second term is entirely due to dimensional changes, whereas the third is
known as the piezoresistive term. The piezoresistive effect is the change in actual
resistivity due to applied strain; it is zero in metals but may be large in some
semiconductors. There are thus two distinct types of strain gauge: metallic types,
with GF ≈ 2 since $\nu \approx 0.5$, and semiconductor types with GF ~ 100. Unfortunately semiconductor devices also have a large temperature coefficient (being similar
to thermistors) so that special temperature-compensation techniques must be used.
Four devices are often used in a bridge arrangement, two being positioned in
places of zero strain, to balance the effects of temperature. However, semiconductor devices are finding increased application, since a film of the material can

often be deposited as an integral part of some other transducer – for example, in pressure or force measurement.

Electrical displacement transducers

These form a large and important class of devices in which a mechanical displacement changes the value of one of the electrical elements L, C or R, a bridge circuit being used to detect the change.

Resistive displacement transducers are simply rotary or linear potentiometers, as shown in figure 3.9.

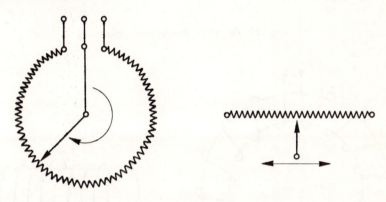

Figure 3.9 *Resistive displacement transducers*

The capacitance C of a parallel plate capacitor of area A and plate separation d and containing a dielectric of relative permittivity ϵ is given by

$$C = \frac{\epsilon \epsilon_0 A}{d}$$

It is clear that the capacitance may be modified by changing ϵ, A or d, giving rise to the three basic types of capacitive displacement transducer: variable permittivity, variable area and variable separation. Typical devices are shown in figure 3.10; three-plate transducers are usually used in practice.

The variable-permittivity capacitance transducer is little used; however the equivalent arrangement in inductive transducers is the only method used. The inductance of a toroid of relative permeability μ and area A, containing a coil of n turns and total wire length l, is given by

$$L = \frac{\mu_0 \mu n^2 A}{l}$$

where μ_0 is the permeability of free space ($4\pi \times 10^{-7}$ H/m). Such a device is not usable as a displacement transducer, and an 'opened out' version in the form of a cylinder and coil, as shown in figure 3.11, has to be used. Unfortunately the

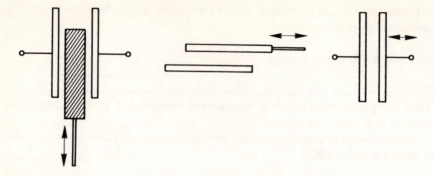

Figure 3.10 *Capacitive displacement transducers*

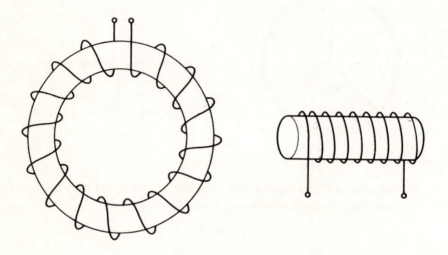

Figure 3.11 *Toroidal and cylindrical inductors*

inductance is not easily calculable since the flux is no longer restricted to the magnetic material; the same formula applies but the effective μ is greatly reduced.

Referring to the above formula, the only way that the device can be used to measure displacement is to vary the effective μ, since n, A and l cannot easily be changed. There are three important types: variable coupling in which the relative inductance of two coils is varied, the differential transformer in which the coupling between a primary winding and two secondaries is changed by the core, and variable reluctance in which the reluctance of a magnetic circuit is changed by a thin magnetic disc.

Optical displacement transducers

These devices are radiant–(mechanical)–radiant modulators, in which an input radiant energy is modulated by a mechanical motion. The best-known examples are angular encoders, in which a pattern of opaque and transparent sections on a disc is rotated and the resulting change in intensity detected by an array of photocells. In absolute encoders, the position of the disc is given by a binary code derived from the photocell outputs, whereas in incremental encoders the number of pulses produced by a given movement is counted.

Grating displacement transducers, sometimes called moiré fringe systems, are similar in principle and comprise two similar patterns on a fixed and a moving disc, with a photocell counting the number of intensity cycles.

3.3.3 Thermal modulators

Thermoresistive transducers

Most thermal modulators are devices whose resistance changes in response to temperature. Both metallic and semiconductor sensors are widely used, though their characteristics differ greatly. The energy level diagrams for a metal and a semiconductor are shown in figure 3.12.

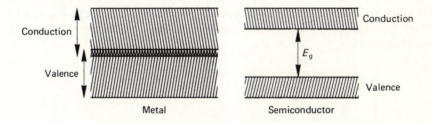

Figure 3.12 *Conduction and valence bands in metals and semiconductors*

In metals the valence and conduction bands overlap, so there are always some conduction electrons and the material has substantial conductivity. In semiconductors there is a gap between the two bands, and the number of electrons N in the conduction band for a given gap depends on temperature T, according to equation (3.4) above.

The resistivity of most materials can be written as

$$\rho_T = \rho_\infty \exp\left(\beta/T\right)$$

where ρ_T is the resistivity at temperature T, ρ_∞ that at very high temperature and β a constant proportional to the energy gap. Considering two temperatures T_1

and T_0, we can eliminate ρ_∞ and write the bulk resistance R as

$$R_{T_1} = R_{T_0} \exp\left[\beta\left(\frac{1}{T_1} - \frac{1}{T_0}\right)\right] \qquad (3.7)$$

For semiconductors β is relatively large and positive, typically several thousand. The change of resistance with temperature is inherently exponential which is a disadvantage, though techniques exist for linearising the response. Most devices are in the shape of beads, bars or discs, and are composed of oxides of nickel, cobalt or manganese. They are known as negative temperature-coefficient (NTC) thermistors, since the slope of the curve is negative. It is possible to obtain a positive slope (PTC devices) over a limited temperature range by suitable doping, but the exact response varies considerably from device to device.

In the case of metals, where an overlap occurs between the valence and conduction bands, the constant β can be considered to be small and negative so the exponential in equation (3.7) can be expanded. The resistance temperature coefficient a_T is given by

$$a_T = \frac{1}{R}\frac{dR}{dT}$$

and is approximately constant, so we obtain the familiar formula for resistance as a function of temperature, that is

$$R_T \approx R_{T_0}\left[1 + a\left(T_1 - T_0\right)\right]$$

The curve has a small positive slope, depending on the particular metal. Figure 3.13 shows the change in relative resistance with temperature for both metals and semiconductors.

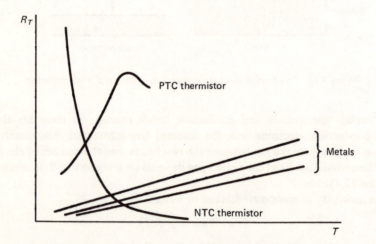

Figure 3.13 *Variation of resistance with temperature for metals and thermistors*

p–n junction devices

It is found that when a silicon diode is forward-biased and carrying a constant current, the temperature coefficient of the voltage drop across it is approximately -2 mV/°C. The exact value varies between individual devices, so calibration is necessary, but the relation is essentially linear (unlike thermistors) and the response time is short. Such devices are particularly cheap of course.

3.3.4   Magnetic modulators: the magnetoresistive and Hall effects

The magnetoresistive effect is the change in resistance of a semiconductor material when subjected to a magnetic field. It is closely associated with the Hall effect, which will be discussed first.

The laws of electromagnetic induction discussed above for magnetic–electrical self-generators and tachogenerators also apply to the movement of individual charge carriers in a magnetic field. When a flat conductor carrying a current i is placed in a magnetic field of induction B normal to its surface, as shown in figure 3.14, an e.m.f. e is produced across the width of the conductor.

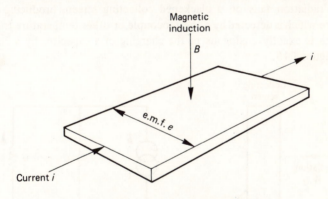

Figure 3.14   *The Hall effect*

For a conductor of thickness t, the e.m.f. is given by $e = K_H Bi/t$ where K_H is the Hall coefficient, dependent on the product of the charge mobility and resistivity of the conductor. The effect is negligibly small in most metals (which have low resistivity) and insulators (which have low mobility), but is appreciable in some semiconductors. Silicon and germanium can be used, but have fairly high resistivity, but indium antimonide is widely used, having $K_H \approx 20$ V/T.

Hall effect transducers may be used for the measurement of current (when they are electrical–electrical modifiers) or magnetic field (in which case the action is that of an electrical–magnetic–electrical modulator). The e.m.f.s produced are only a few microvolts and the effect is strongly temperature-dependent, so the detectivities obtained are low. The related magnetoresistive effect arises when the Hall voltage is short-circuited; the charge carriers are then deflected, resulting in an increased path length and consequently increased resistance. This effect is not much used at present, but its importance will increase as solid-state sensors are developed.

3.4 Modifiers

Since there are six basic forms of energy there should be six modifiers, having the same form of energy at input and output. Most modifiers convert energy between the two variables in the system — for example, force–velocity, flow–pressure or radiation–temperature. However, the only large groups of devices are the mechanical and radiant (or thermal) modifiers.

3.4.1 Radiant modifiers

Radiant modifiers are used in thermal radiation detectors in which an input stream of radiation falls on a blackened collecting screen, producing a rise in temperature which is detected by a thermocouple or other temperature transducer. The process is exactly analogous to the charging of a capacitor by a current, as shown in figure 3.15.

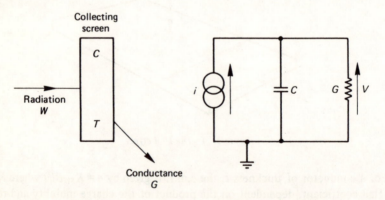

Figure 3.15 *Radiant modifier and its electrical equivalent*

The temperature rise is given by

$$\delta T = \frac{\epsilon W/G}{(1 + j\omega\tau)}$$

where ϵ is the emissivity of the screen and $\tau\,(= C/G)$ the thermal constant.

A rather similar form of thermal modifier is used in anemometers, in which a flow of fluid changes the temperature of a heated wire, transducing heat flow into temperature rise.

3.4.2 *Mechanical modifiers*

Mechanical modifiers convert a mechanical input into another mechanical form, usually the other variable. One large class of devices is elastic elements, usually used for force measurement, in which the force is applied to a spring, cantilever beam or bar, producing a deflection (which is measured by a second transducer). The dimensions of the device are chosen to suit the range of forces of interest, and the displacement per unit force is easily calculable from the appropriate modulus of the material used. It is often convenient to measure strain rather than deflection, which is calculated from the well-known equations for a stressed bar.

Another important class of modifiers is pressure-sensitive components, such as tubes and bellows. Much ingenuity has been dissipated by engineers in designing hollow tubes of various shape (C-type, spiral, helical etc.) but such devices are rather crude in principle and often unsatisfactory in operation, and one cannot help wishing they had not bothered. A stretched diaphragm or thin plate that is deflected by the applied pressure is more satisfactory, the displacement being conveniently detected by capacitive or inductive transducers. These devices are linear over only a small range of deflections (typically about half the thickness of the diaphragm), and corrugated diaphragms are sometimes used to give a greater range.

A third class of mechanical modifiers is used for flow measurement. The principle mostly employed is that of providing some restriction to the flow and measuring the corresponding pressure drop, in analogy to measuring an electric current by finding the voltage across a small series resistor. Such devices are not very satisfactory, since the flow varies across the tube and may be streamline (without eddies etc.) or turbulent, depending on the appropriate Reynolds number (a function of velocity, diameter, density and viscosity); in addition, the flow rate is usually proportional to the square root of the pressure drop.

In recent years there has been a lot of interest in resonant sensor systems, in which the resonant frequency of a mechanical structure varies with a parameter such as pressure or force. Such devices are also mechanical modifiers, and will be discussed in chapter 9, which deals with recent developments and future trends.

3.5 Exercises on chapter 3

3.5.1. (a) Explain how transducers can be classified on the basis of energy conversion into modifiers, modulators and self-generators, explaining the essential difference between each type.

(b) List the most important types of self-generators (which produce an electrical output), the most important modulators (having electrical excitation and providing an electrical output), and the principal modifiers, in each case stating the physical principle involved and giving an example of a transducer employing that principle.

3.5.2. An alternative method of classifying transducers is into 'transforming transducers', in which the action is a transduction between corresponding variables (through/across) and 'gyrating transducers', in which a 'through' variable is transduced into an 'across' variable and vice versa. Classify the following transducers on this basis.

(i) The main types of self-generator (photovoltaic, pyro-electric, electro-magnetic, piezo-electric, thermo-electric).

(ii) The main types of modifer (mechanical, radiant, thermal).

3.5.3. Discuss the various applications of Faraday's law of induction to transducers, stating in each case the precise physical effect and the type of transducer involved (for example, magnetic–electrical self-generator, electrical–magnetic–electrical modulator etc.).

4

Transducer Bridges and Amplifiers

In the previous chapter we discussed transducers for all the main forms of energy except the electrical form. Electrical transducers may be modifiers, self-generators or modulators, of course, but the most important here are those having an electrical output, such as amplifiers and bridge circuits, since these are usually required to follow the transducers already discussed. Both are strictly modulators, since they involve a separate power source which is modulated by the (small) electrical signal. It is necessary to consider these devices at this stage so that we can deduce the responsivities of the transducers to be discussed in the next chapter.

4.1 Transducer bridges

As we have seen above, many transducers are modulators and control the flow of (usually electrical) energy. For example, a thermistor transduces temperature into change in resistance, but to obtain a usable signal a current must be applied and the resulting voltage measured. In general, where the transduction is into an electrical form, such as resistance, capacitance or inductance, some form of bridge circuit is required. Bridges may be excited by direct current, in which case they are always resistive, or by alternating current, when they may be resistive, capacitive or inductive.

4.1.1 D.c. bridges (resistive)

The most general form of resistive bridge is the four-arm arrangement, shown in figure 4.1 together with its equivalent circuit. Any one or all the resistors may actually represent transducers. In measuring temperature one arm may be a thermistor and the others fixed (or variable) resistors; alternatively, in measuring

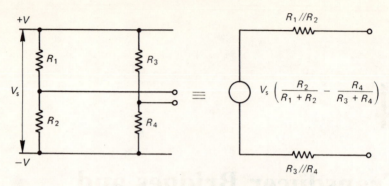

Figure 4.1   *General d.c. resistive bridge and its equivalent circuit*

strain with a cantilever beam, all four arms could be suitably connected strain gauges.

There are three cases of interest.

(i)   Double push-pull: all four arms are identical transducers, with

$$R_1 = R_2 = R_3 = R_4 = R$$

and

$$\delta R_1 = -\delta R_2 = -\delta R_3 = \delta R_4 = \delta R$$

The equivalent circuit reduces to that in figure 4.2(a).

(ii)   Single push-pull: R_1 and R_2 are transducers with

$$R_1 = R_2 = R$$

and

$$\delta R_1 = -\delta R_2 = \delta R$$

R_3 and R_4 are fixed and equal (often equal to R).

The equivalent circuit becomes that in figure 4.2(b).

(iii)   Simple transducer: R_1 is the only transducer. Usually

$$R_1 = R_2 = R_3 = R_4 = R$$

We now have the equivalent circuit of figure 4.2(c).

Note that in figure 4.2(c) the resistance $R_1//R_2$ does not remain constant, unlike cases (a) and (b) where $\delta R_1 = -\delta R_2$. This causes a non-linear output for large changes in R_1.

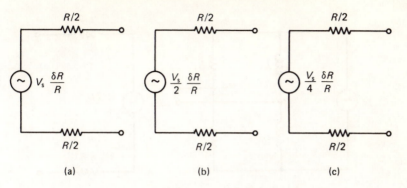

Figure 4.2 *Equivalent circuits of d.c. bridge arrangements*

4.1.2 A.c. bridges

An a.c.-excited resistive bridge could, of course, be obtained by simply replacing the d.c. excitation of figure 4.1 by a suitable oscillator. However, it is generally preferable to use a transformer, as shown in figure 4.3, since this gives better isolation between the oscillator and the bridge, but in particular permits the two sides of the bridge to be exactly equal and opposite in excitation (or any other ratio that one wishes). It is very easy to wind a transformer with identical secondaries, because one simply uses two pieces of wire twisted together. Another advantage is that the bridge requires only two resistors (both of which are usually transducers) and a single-ended output is easily obtained by grounding the centre part of the secondary.

Inductive and capacitive bridges are obtained in the same way. Push–pull types are preferred and the arrangements and equivalent circuits are shown in figure 4.4(a) and (b).

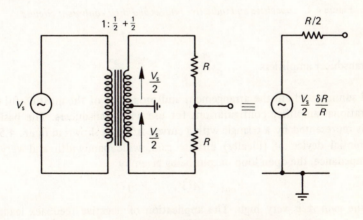

Figure 4.3 *A.c. resistive bridge and its equivalent circuit*

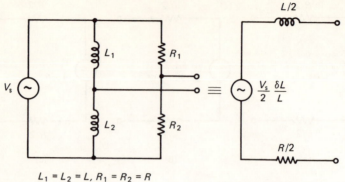

$$L_1 = L_2 = L, R_1 = R_2 = R$$
$$\delta L_1 = -\delta L_2 = \delta L$$

(a)

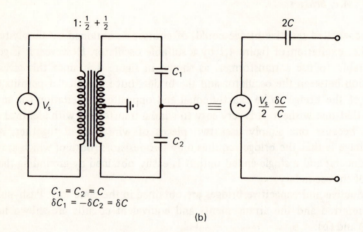

$$C_1 = C_2 = C$$
$$\delta C_1 = -\delta C_2 = \delta C$$

(b)

Figure 4.4 *Capacitive and inductive bridges and their equivalent circuits*

4.2 Transducer amplifiers

We will summarise here the arrangement and properties of the most useful forms of operational amplifier configurations for use with transducers. The basic amplifier is represented by a triangle with a curved front, as shown in figure 4.5. It is a differential device of (ideally) infinite gain, wide bandwidth and very high input impedance, the open loop output being given by

$$V_{out} = A(V_+ - V_-)$$

where the gain A is very high. The application of negative feedback leads to a number of useful arrangements of stable and precisely known gain. In each case

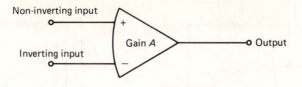

Figure 4.5 *Ideal operational amplifier*

the feedback tends to make the voltage difference between the two input terminals very close to zero.

(i) Inverting amplifier (virtual-earth amplifier)

The most familiar arrangement is the virtual-earth amplifier of figure 4.6(a). Since the non-inverting terminal is grounded the other terminal is forced to remain close to zero, and it is easy to see that the voltage gain is $-R_2/R_1$ and the input impedance R_1.

(ii) Current amplifier

This is similar to the inverting amplifier, except that resistor R_1 is omitted so that the input current flows through R_2. The gain (which has dimensions of volts/ampère) is equal to R_2 and the input impedance is very low (since the current flows into the virtual-earth point).

(iii) Non-inverting amplifier

This is not a virtual-earth arrangement, the voltage of the inverting terminal remaining close to that of the input. The voltage gain can be shown to be positive and of value $(R_1 + R_2)/R_1$. The input impedance is very high (equal to the open loop value).

(iv) Differential amplifier

The differential amplifier is very widely used in transducer applications. It is a combination of the inverting and the non-inverting amplifiers, and has gain $+ R_2/R_1$ via the two terminals and input resistance R_1. It is not a virtual-earth amplifier.

(v) Charge amplifier

This arrangement is used with capacitive transducers and has the advantage that since the feedback is capacitive the response with respect to the transducer

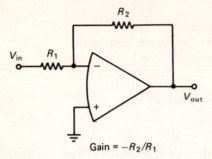

Gain = $-R_2/R_1$

(a) Inverting amplifier

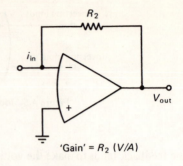

'Gain' = R_2 (V/A)

(b) Current amplifier

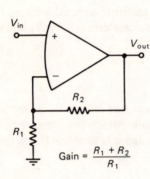

Gain = $\dfrac{R_1 + R_2}{R_1}$

(c) Non-inverting amplifier

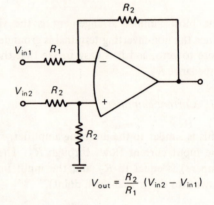

$V_{out} = \dfrac{R_2}{R_1}\ (V_{in2} - V_{in1})$

(d) Differential amplifier

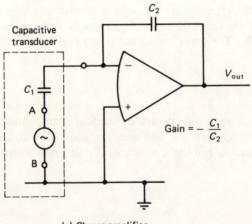

Gain = $-\dfrac{C_1}{C_2}$

(e) Charge amplifier

Figure 4.6 *Basic operational amplifier arrangements*

terminals A and B is independent of frequency. The gain is given by the ratio of the capacitances and is $-C_1/C_2$, and the input impedance is very low.

The name is derived from applications where charge is being measured, and is rather inappropriate in this application. It is really a capacitance-to-voltage converter.

4.3   Exercises on chapter 4

4.3.1. Draw an equivalent circuit of the system of figure 4.7. Find the overall responsivity assuming that the thermistor has a resistance of 10 kΩ at the temperature of operation, when its curve of resistance against temperature has a slope of 0.5 kΩ per °C.

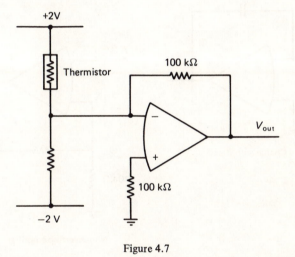

Figure 4.7

4.3.2. A differential capacitive displacement transducer comprises two capacitances in a bridge arrangement. The capacitances are each of 10 pF when in the zero position, and change by ±1 pF per mm. The output may be connected to either a charge amplifier or a non-inverting amplifier, as shown in figure 4.8. Draw an equivalent circuit and find the overall responsivity in each case.

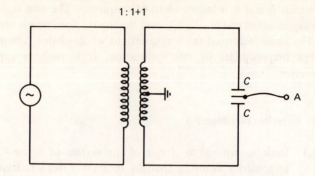

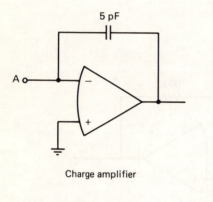

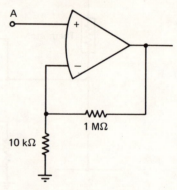

Figure 4.8

5

Basic Transducers for Length

In chapter 3 we discussed the classification of transducers and gave a comprehensive summary of the useful physical effects available. Although such overviews may be valuable and interesting, they are not very helpful in designing or selecting a device for a specific application. The remainder of the book will therefore be devoted to practical aspects. The three most important measurement areas are length, temperature and radiation; in each case we will discuss the basic theory where appropriate, but will concentrate on the design and operation of the most useful devices.

5.1 Classification of length transducers

The standard of length was originally the metre bar, but this was replaced in the 1960s by a definition involving the wavelength of a Krypton discharge lamp. It has been redefined recently (1983) in terms of the velocity of light and the standard of time, and is the length of the path traversed by a beam of light in vacuum in 1/299,792,458 seconds.

We will include under length transducers all length-related quantities, such as displacement, velocity, acceleration and also strain. We will mostly discuss translational devices, mentioning rotational devices only where appropriate. Transducers may be used for relative or absolute measurements. For example, we may wish to detect the relative motion between two objects or we may wish to measure the 'absolute' vibration of an object (for example, in a rocket or spacecraft) where no reference surface is available. However, we will consider only relative transducers here.

5.2 Displacement transducers

The most important applications for displacement transducers involve ranges from a few micrometres to a few metres. Responsivities vary from about 1 V/m to 10^5 V/m or more, with detectivities as high as 10^{12}/m. The two main classes of displacement transducers are electrical and optical. The former involve one of the primary electrical elements $(R, C$ or $L)$ and the latter mostly involve some form of opaque/transparent pattern on a disc or plate and are often digital.

5.2.1 Electrical displacement transducers

There is an important general formula for the responsivity of a linear electrical displacement transducer. If the impedance of the device varies from zero to Z_{max} over the range l_{max}, and the excitation voltage is V_{ex}, then for a change δZ corresponding to a displacement δl we have an output δV given by

$$\frac{\delta V}{V_{ex}} = \frac{\delta l}{l_{max}} = \frac{\delta Z}{Z_{max}} \quad \text{and} \quad r = \frac{V_{ex}}{l_{max}} \qquad (5.1)$$

This can easily be seen to apply to a resistive potentiometer, as shown in figure 5.1.

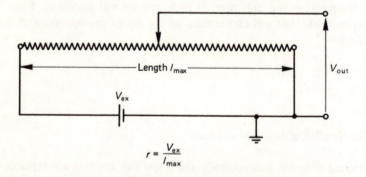

$$r = \frac{V_{ex}}{l_{max}}$$

Figure 5.1 *Responsivity of resistive potentiometer*

It can be seen from equation (5.1) that one can obtain a high value of r for a given excitation only at the expense of the range, and while this may now be obvious it has often not been realised in the literature.

Resistive displacement transducers

Resistive transducers may be translational or rotational, the simple rotary potentiometer being the most widely used type. They are relatively cheap and may have linearity of 0.1 per cent or better. Their main disadvantages are associated with the necessity of physical contact between the wiper and winding, and some friction and wear are inevitable. In cheap wire-wound devices the gauge of the wire may limit the resolution, but this can be avoided by a suitable design of wiper. A comprehensive treatment is given by Doebelin (1966).

Devices are available with ranges up to about 50 cm and rotational types up to 20 or more turns. However, although cheap and useful in simple applications, resistive transducers are unsuitable for most precision applications.

Capacitive displacement transducers

We saw in chapter 3 that the most important capacitive devices are the variable-area and variable-separation transducers. It is not usually convenient to employ only two plates, and differential arrangements with three plates are always used.

(i) *Variable-area type* A simple variable-area transducer is shown in figure 5.2.

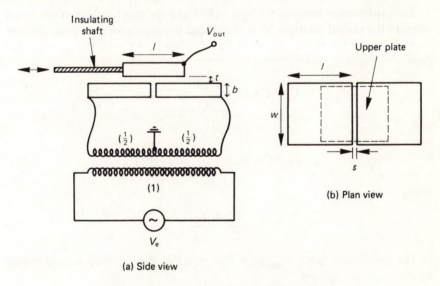

Figure 5.2 *Variable-area capacitive transducer*

The device comprises three identical plates, two of them fixed and separated by a small distance s, and the upper one capable of sliding over the fixed plates with a constant separation t. The two fixed plates are excited in antiphase by means of a centre-tapped transformer $(1 : \frac{1}{2} + \frac{1}{2})$. It is clear that when the centre plate is symmetrically placed with respect to the fixed plates no net voltage will be induced on it, whereas movement to the left will produce an increasing voltage of phase the same as that of the left-hand fixed plate, and similarly for movement to the right. The magnitude indicates the total displacement and the phase the sense. The range is clearly equal to the plate length l.

The equivalent circuit of the arrangement is shown in figure 5.3.

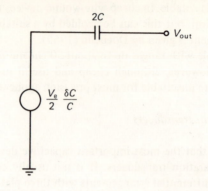

Figure 5.3   *Equivalent circuit of figure 5.2(a)*

The capacitances between the upper plate and the two fixed plates are each C when in the central position. δC is the change in capacitance for a small displacement δx, when the two capacitances then become $C_1 = C + \delta C$ and $C_2 = C - \delta C$. Now

$$C_1 = C + \delta C = \frac{\epsilon \epsilon_0 w}{t} \, (l/2 + \delta x)$$

and

$$C_2 = C - \delta C = \frac{\epsilon \epsilon_0 w}{t} \, (l/2 - \delta x)$$

giving

$$\frac{\delta C}{C} = \frac{\delta x}{l/2}$$

The responsivity is $r = V_{out}/\delta x = V_e/l$, equal to the excitation voltage divided by the range, as expected.

It is fairly easy to construct such a transducer. The plates can be 2–3 cm square pieces (2.54 cm is a good number), cut from 3 mm brass sheet. The distance s is not critical but should be 1 mm or less, and the separation t should be about 1 mm. This can be arranged by means of an insulated sheet, fixed to the two lower plates, which can contain guides to ensure correct movement of the upper plate. An alternative method is to use an arrangement of springs, as shown in figure 5.4, which provides smooth parallel motion (though the separation t changes slightly for large deflections).

The transformer is particularly easy to construct. Three 10 m lengths of thin copper wire are twisted together with a hand drill and wound together on a small (1 cm ID) pot core. This ensures that the secondaries are identical but produces a

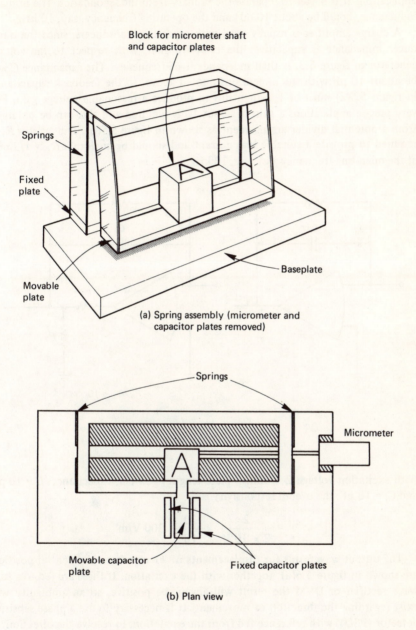

Figure 5.4 *Parallel-motion spring mounting (with variable separation transducer)*

1:1 + 1 transformer rather than the $1:\frac{1}{2} + \frac{1}{2}$ described above. For very precise applications it is better to separate the primary from the secondaries. The primary inductance should be about 10 mH and the operating frequency, say, 10 kHz.

A charge amplifier is usually used with capacitive transducers, since the transducer impedance is capacitive; the effective gain, with respect to the voltage generator in figure 5.3, is then independent of frequency. The capacitance C will be about 10 pF with the dimensions quoted above, so the feedback capacitance in figure 5.5(a) must be less than this to produce a reasonable voltage gain. For very precise applications a gain of about 10 is required and this can be obtained from a potential divider arrangement as shown in figure 5.5(b). The resistor R_f is required to provide a suitable bias current and should be such that $R_f \gg 1/2\pi f C_f$ at the operating frequency f of, say, 10 kHz.

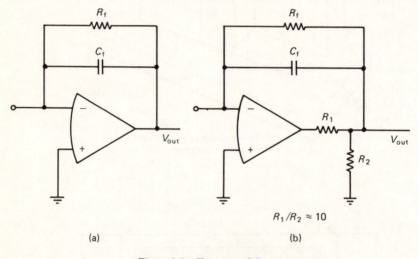

$R_1/R_2 \approx 10$

(a)   (b)

Figure 5.5 *Charge amplifiers*

With excitation voltage 2 V r.m.s., plate length l = 2 cm, capacitances C = 10 pF and C_f = 10 pF, the overall responsivity is

$$r = \frac{2}{2 \times 10^{-2}} \times \frac{20}{10} = 200 \text{ V/m}$$

The output waveforms for displacements of ±1 mm from the central position are shown in figure 5.6(a), together with the excitation. If these are fed to a full-wave rectifier or DVM the result will always be positive, so an ambiguity will exist regarding the direction of movement. It is necessary to use a phase sensitive detector (PSD), with reference fed from the excitation, to resolve the direction. A PSD is essentially a multiplier so in-phase reference and excitation produce a positive result and antiphase signals a negative result. The output as a function of displacement is shown in figure 5.6(b).

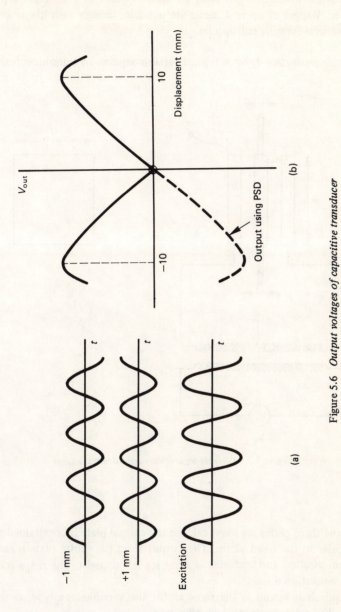

Figure 5.6 *Output voltages of capacitive transducer*

The range of the transducer (equal to the plate length) is 2 cm. However, much larger ranges are possible by using a multiplate arrangement, in which several adjacent plates (instead of just two) are each supplied by a different tap on the transformer. Ranges of up to a metre are possible, though with low responsivity since the general formula still applies.

(ii) Variable-separation type A typical variable-separation transducer is shown in figure 5.7.

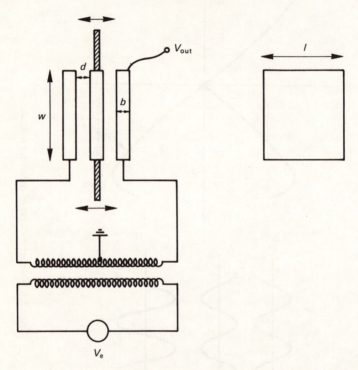

Figure 5.7   *Variable-separation capacitive transducer*

Again the three plates are identical, but the central plate is constrained to move perpendicular to the fixed plates. The output from the centre plate is zero when in a central position and increases as it moves left or right. The range is equal to twice the separation d.

The equivalent circuit is the same as for the variable-area type, as shown in figure 5.3. For a displacement d we now have

$$C_1 - C_2 = 2\delta C = \frac{\epsilon\epsilon_0 lw}{d - \delta d} - \frac{\epsilon\epsilon_0 lw}{d + \delta d} = \frac{\epsilon\epsilon_0 lw}{d^2 + \delta d^2} \times 2\delta d$$

and

$$C_1 + C_2 = 2C = \frac{\epsilon\epsilon_0 lw}{d^2 + \delta d^2} \times 2d$$

giving

$$\frac{\delta C}{C} = \frac{\delta d}{d}$$

The responsivity is thus $V_e/2d$, as expected from the general formula. It is interesting to note that the device is inherently linear, although a two-plate version would be very non-linear. However, as shown below, non-linearity is usually observed in practice.

Constructional details of this transducer are similar to those of the variable-area device. The separation d should be about 1 mm to produce a reasonable capacitance of about 10 pF, so the range is very small. Mechanical guides can be made to restrict the movement of the centre plate, but the spring arrangement of figure 5.4 is very convenient.

The transformer and amplifier are similar to those above. However, from the equivalent circuit and amplifier together, it is clear that the output from the amplifier is

$$V_{out} = \frac{V_e}{2} \times \frac{\delta d}{d} \times \frac{C}{C_f} \qquad \text{so} \qquad r = \frac{V_e}{2} \times \frac{C}{dC_f} \approx 2000 \text{ V/m}$$

Unfortunately C is not constant, being equal to $\epsilon\epsilon_0 lwd/(d^2 + \delta d^2)$, so the output shows a serious non-linearity. This is because a current (low input impedance) amplifier is used. If a voltage (high input impedance) type was used, one would obtain a linear output. However, the low impedance arrangement has considerable practical applications in insensitivity to stray capacitances and electromagnetic interference, so it is always used in practice. The small range limits the application of the transducer, and it is usually used as a null-sensing device, so that the linearity is of little importance. The form of the output against displacement is as shown in figure 5.8, again using a PSD to obtain a bipolar output, indicating direction of motion.

The slope in the central region is that deduced above (2000 V/m).

Inductive displacement transducers

Inductive transducers are widely used in industry, since they are compact and robust. They are less easily affected by environmental factors (humidity, dust etc.) than capacitive types. Their detectivity is adequate for most applications, though capacitive types are preferable for very precise measurements, since their detectivity is higher and the (unwanted) forces exerted in the measurement are much lower.

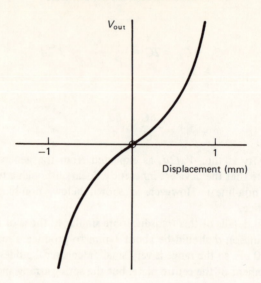

Figure 5.8 *Output voltage of variable-separation transducer (with PSD)*

(i)   Variable-coupling transducers These transducers consist of a former holding a centre-tapped coil, and a ferromagnetic plunger, as shown in figure 5.9.

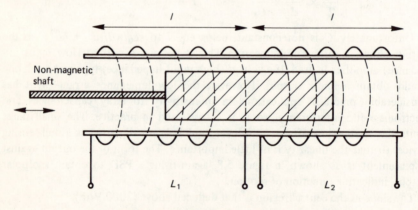

Figure 5.9 *Variable-coupling transducer*

The plunger and both coils have the same length l. As the plunger is moved the inductances of the coils change; ideally the maximum inductances will be very high and the minimum (when the plunger is in the other coil) quite low. The range is therefore l. The two inductances L_1 and L_2 are placed in a bridge circuit with two equal balancing resistors R, followed by an amplifier and PSD, as shown in figure 5.10.

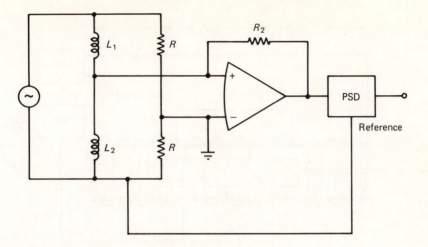

Figure 5.10 *Inductive bridge*

If the inductances in the central position are each L, corresponding to a plunger length $l/2$, a change in displacement of δl produces changes $+\delta L$ and $-\delta L$, so ideally $\delta L/L = \delta l/(l/2)$, and the output from the bridge is $(V_{ex}/2 \times \delta L/L)$ giving responsivity $V_{ex}/$range as usual.

It is particularly easy to construct a transducer of this type. One simply winds a centre-tapped coil on a suitable former, using a rod of mild steel for the plunger. Devices, sometimes known as linear displacement transducers or LDTs, are available commercially up to about 0.5 m in range.

(ii) Differential transformers These are probably the most widely available displacement transducer, being very robust yet capable of detecting displacements in the nanometre range. They comprise a transformer in which the coupling between primary and a split secondary depends on the position of a ferromagnetic plunger, as shown in figure 5.11. The two halves of the secondary winding are connected in series opposition.

There are various designs, but in figure 5.11 the coils all have the same length l, but the plunger has length $2l$. In its central position it protudes equally into the two secondaries, and in its extreme positions produces maximal coupling with one secondary and minimal with the other so that the range is l. Since the secondaries are in opposition the output is zero in the central position and increases on either side (but with opposite phase).

If we assume that the coupling is perfect in the two extreme positions, then for a unity turns ratio transformer (that is, $1 : \frac{1}{2} + \frac{1}{2}$) the secondary voltage will be equal to half the excitation voltage in these positions, and the responsivity is therefore V_{ex}/l.

Commercially produced devices have ranges from about 1 mm to 20 cm and

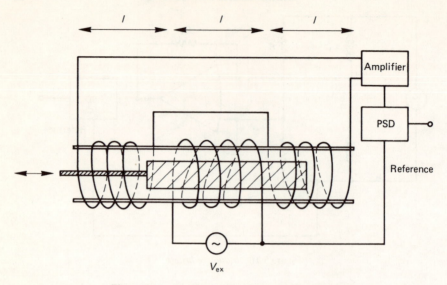

Figure 5.11 *Linear variable differential transformer*

are usually known as LVDTs (linear variable differential transformers). The windings are enclosed in a stainless steel sheath, as illustrated in figure 5.12, with four wires protruding. Two wires are simply connected to an oscillator at, say, 1 kHz and the other two into an amplifier, so the device is very easy to use. The responsivity (unfortunately often called 'sensitivity') is usually given in the form

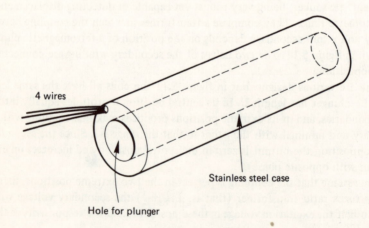

Figure 5.12 *External view of LVDT*

of millivolts per volt excitation per millimetre displacement. The turns ratio is not usually given, so one cannot apply the general formula for responsivity.

It is fairly easy to construct an LVDT, by simply dividing a former into three sections, winding three similar coils, and connecting the secondaries in opposition. Unfortunately such a device will almost certainly have one outstanding characteristic; it will be a very poor LVDT! The problem is that it is not easy to make the two halves of the secondary identical, and their inductance, resistance and capacitance will be different, causing a large unwanted quadrature output in the balance position. Although this can be rejected by a PSD it may seriously limit the usable amplifier gain, and may change with temperature so that attempts to balance it out are hazardous. This is an interesting contrast to the capacitive transducers, where it is easy to make an excellent centre-tapped transformer because the windings may be twisted together. The fact that the two halves are physically separated in the LVDT is an inherent weakness of the device and precision coil-winding equipment is required to reduce the problem to an acceptable value.

(iii) Variable-reluctance transducers These devices can be thought of as the inductive equivalent of the variable-separation capacitive transducer. They have a very small range and are used in rather specialised applications such as pressure transducers. They consist essentially of two closely spaced coils with a thin ferromagnetic disc between them, the position of which changes the inductance of the coils. The disc can be thought of as changing the reluctance of the associated magnetic flux circuit, as illustrated in figure 5.13.

The transducer is operated in a similar way to the variable-coupling type, the responsivity being given by $V_{ex}/2d$, which will be very high since d is usually about 1 mm. Unfortunately the magnetic forces imposed on the disc are quite large, and this limits the application severely. However, the disc can conveniently be a thin diaphragm, hence the application to pressure measurement.

Rotational electrical transducers

Most of the resistive, capacitive and inductive transducers described above are available in rotational form, so the same basic theory applies. There are a few additional rotary devices of interest, notably synchros, which are much beloved of control engineers. They consist of a cylindrical coil which can rotate about an axis, as shown in figure 5.14. It is excited by an a.c. voltage, and surrounded by three coils equally spaced at intervals of 120°. At a given input angle of the central coil, the outer coils detect three voltages which uniquely define the input angle. Synchros have a rather serious disadvantage (this has also been said of control engineers) in that they employ slip rings to feed the central coil, and although convenient in many cases they are unsuitable for high precision measurement.

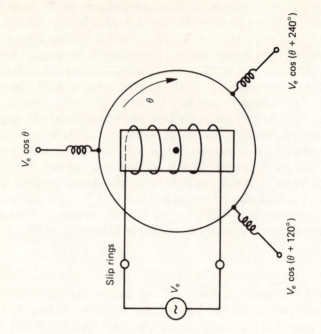

Figure 5.14 *Synchro (schematic)*

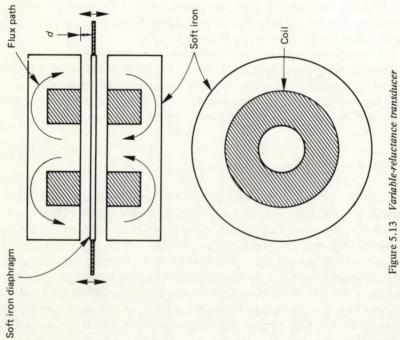

Figure 5.13 *Variable-reluctance transducer*

5.2.2   *Optical transducers*

Optical transducers have become very popular in recent years, partly because they are mostly inherently digital and partly because they are fairly immune from electrical interference. The main types are encoders, carrying some form of code representing position, and gratings. Encoders may be absolute or incremental, and are usually used for angular measurements, whereas gratings are purely incremental and are used for both translation and rotation.

Absolute angular encoders

A binary-coded absolute encoder is shown in figure 5.15(a). It consists of a number of concentric tracks containing a pattern of opaque and transparent sections such that a unique binary code can be read from the patterns for any angle. The patterns may be read by means of photocells, as shown in figure 5.15(b).

One problem with this arrangement, to which some books gleefully devote many pages, is that when several tracks change together (as between positions 3 and 4 in figure 5.15(a)), a serious error could occur if a readout is attempted midway between these positions. The well-known Gray code (figure 5.15(c)) was developed to overcome this, and has the property that only one track changes at a time so whenever the readout occurs it cannot be in error by more than one bit. Simple algorithms exist for converting between Gray code and binary.

Absolute encoders are available with 8, 10 or 12 tracks, but become increasingly expensive, and also physically large, as the number of tracks increases.

The need for the Gray code arose because the readout was essentially asynchronous; it could be required at any time and was not related to any reference or clock waveform. However, recent technological developments have made it usual, or at least convenient, to read the outputs of encoders directly via microprocessors, and the earlier difficulties no longer apply. One can either reject an invalid reading occurring at a crossover point (as one would do intuitively) or use an additional clock track to ensure that readings are taken only where valid (the previous value is taken if the disc comes to rest at a crossover on the clock track).

Actually the complication of requiring n tracks for a measurement to an accuracy of n bits is no longer necessary. A rather novel alternative is to use a single track containing a maximal-length binary sequence of, say, 256 digits for 8 bits. Such a sequence contains all the different combinations of 8 digits exactly once each, and every angular readout (determined by 8 adjacent bits) is unique. A separate clock track is usually used, as in figure 5.16 which shows a simple 4-bit version. It is interesting to note that the eight photocells do not have to be adjacent; any equal spacing is satisfactory. In fact one can actually use only two

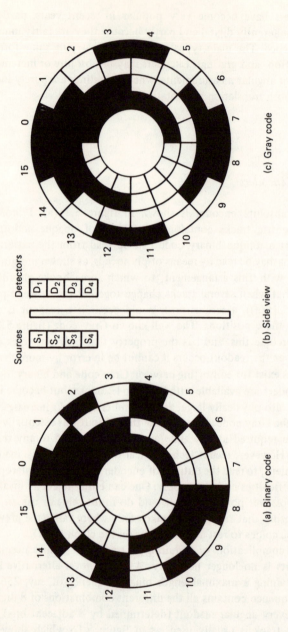

Figure 5.15 *Absolute angular encoders*

photocells, reading them serially into the processor, but this requires an initial movement over eight digits to define the first position. A few extra bits can be obtained by interpolation between the clock track digits.

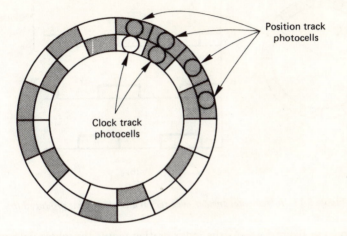

Figure 5.16 *Absolute encoder with maximal-length binary sequence (note that this sequence includes the 0000 state)*

Incremental angular encoders

Incremental encoders comprise a disc with a single binary track; this enables them to be made in 'slotted' form rather than transparent/opaque, as for coded discs. The slots are detected by a photocell arrangement feeding a counting system, so the resolution is essentially equal to the angle between the slots. In order to detect direction of rotation two photocells are placed about a quarter of a slot width apart and the relative phases of the two signals reverses when the direction changes. The signals are fed to a direction-detector logic circuit which controls an up/down counter. Typical discs usually have numbers of slots between about 48 and 96. Figure 5.17 shows a disc and the two photocell signals.

Optical gratings

A grating measurement system can be thought of as a development of a well-known mechanical–optical modulating transducer which we have omitted to mention so far — the ruler. One could observe a grating containing many lines with a microscope and simply count lines, but this would be a little tedious; unfortunately the process is not easy to automate. Grating measurement systems overcome the problem by using two identical equal-mark/space gratings, one fixed (the scale grating) and the other (the index grating) capable of moving over

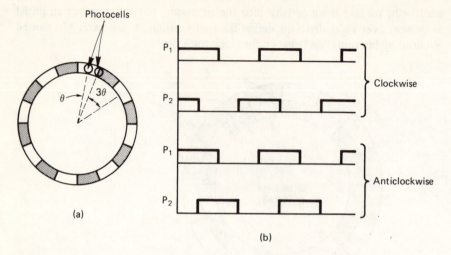

Figure 5.17 *Incremental angular encoder (a) and waveforms produced (b)*

it, as shown in figure 5.18. As the index grating moves the whole field goes light or dark, so that a large photocell can be used, covering many lines, unlike the situation with the ruler where individual lines have to be observed. This has the added advantage that errors in single lines are automatically averaged out. These systems originally used two gratings with a small angle between them. This produced

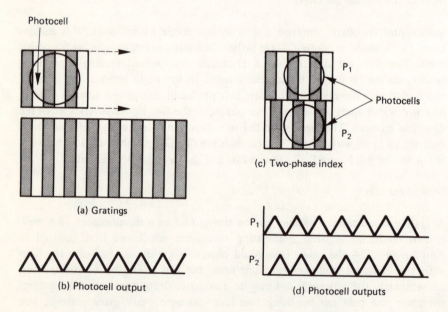

Figure 5.18 *Optical gratings and waveforms produced*

a geometrical pattern known as a Moiré fringe, named after the moiré silk material used in nightdresses. A study of such patterns is strongly recommended.

With a single photocell feeding a counting system there would be no indication of direction of movement. This is usually solved by using a two-phase index, containing two sets of lines displaced by a quarter of a line spacing. Quadrature signals are produced, as in figure 5.17, feeding a direction detector and an up/down counter.

Gratings are available fairly cheaply in a wide range of sizes, for both translation and rotation. The resolution of the measurement is essentially equal to the spacing between the lines, of course, but it is possible to interpolate electronically by a factor of 10 or even 100. This entails essentially finding the ratio of light transmitted in a given position, to the maximum transmitted when coincident, but is actually done by a phase-shifting technique and a set of comparators. This type of electronics is relatively cheap now, so fairly coarse gratings (say 100 lines/cm) are often used in practice followed by suitable interpolation. Difficulties arise in setting up finer gratings, since diffraction in the light falling on the first grating produces a loss in signal, and the gratings have to be placed very close together.

Construction of optical transducers

Most optical transducers are produced photographically, though this may be followed by etching for incremental encoders. The master encoders or gratings are produced by machining, which is of course expensive. However, the availability of microcomputers has made it relatively easy to construct accurate devices, by producing a master on a graph plotter. It is easy to program, say, a BBC or an Apple microcomputer to draw an A4 size grating via a graph plotter, with an effective resolution of better than 1 mm, and this can then be reduced photographically to the required size. Incremental encoders and radial gratings can be produced very easily, and Gray code (or pseudo-random codes) with a little more difficulty (in programming).

5.3 Velocity transducers

Velocity transducers employ the electrodynamic effect, and are therefore self-generating. There are two main types — moving coil and moving magnet. The former are used mainly in instruments such as accelerometers and velocity meters, whereas the latter are available as off-the-shelf transducers and are very similar to LVDTs.

(i) Moving coil transducers The most usual arrangement comprises a cylindrical magnet in a ferromagnetic yoke with pole-pieces, as shown in figure 5.19.

The coil is usually wound on a former and attached to the object whose velocity is required. The responsivity is given by

$$r = Bl \ \text{V/ms}^{-1}$$

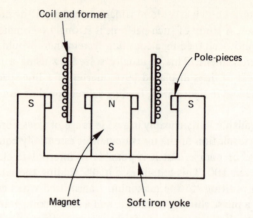

Figure 5.19 *Moving coil transducer*

where B is the induction and l the length of wire cut by the flux. In order to obtain a high value we require both B and l to be large; the air gap has to be small to make B large, so many turns of thin wire are necessary, and in some cases a self-supporting coil (that is, with no former) is used. It is very easy to obtain a responsivity of, say, 1 V/ms^{-1} but in order to obtain a high value (say, 100 V/ms^{-1}) a magnet of about 5 cm diameter and about 10,000 turns of thin wire is necessary. It is not very easy to calculate the responsivity theoretically, but it can be done approximately since for most magnets with air gaps of up to about 5 mm the induction B is about 1 Wb/m^2 and, by calculating the length of wire cut, one gets a reasonable estimate of the responsivity. For example, with a magnet diameter of, say, 1 cm, a coil of diameter 1.5 cm and pole-pieces such that 100 turns are cut, the effective value of 1 is

$$2\pi \times \frac{1.5}{2} \times 10^{-2} \times 100$$

so $r \approx 5$ V/ms^{-1} (assuming $B = 1$ Wb/m^2).

As pointed out in chapter 3, the electrodynamic effect is reversible, a current i producing a force Bl Newtons, and this can be used to calibrate the transducer. Since the device is usually used with accelerometers in any case, one simply applies a known force to the instrument by adding a mass and balances this by applying a current to the coil, giving $Bl = mg$. The process can of course be done on a beam balance if required.

(ii) Moving magnet transducers The simplest form of moving magnet transducer is just a single coil and magnet, as shown in figure 5.20(a). This has the disadvantage that the responsivity varies sharply with the exact position of the magnet. In fact, if the magnet was entirely within the coil there would be no net flux cut and

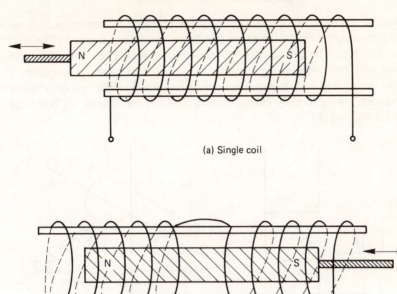

(a) Single coil

(b) Push-pull

Figure 5.20 *Moving magnet transducers*

no output. A better arrangement is the push–pull device, sometimes known as an LVT (linear velocity transducer), shown in figure 5.20(b), in which the second half of the winding is reversed in sense. This has a much wider range of position over which the responsivity is constant.

The appearance of this push–pull arrangement is very like that of an LVDT. In fact one can use an LVDT as an LVT by simply applying a d.c. voltage to the primary, which magnetises the ferromagnetic core, taking the output from the (series opposition) secondary.

Rotational velocity transducers

The rotational equivalent of the moving coil velocity transducer is the tachometer. This is another device favoured by control engineers. There are two main types — the d.c. tacho and the a.c. tacho — though both have disadvantages. The d.c. tacho is essentially a d.c. generator with permanent magnet excitation. A coil rotates in the field of a magnet, many poles being used to produce a reasonably smooth output, and slip rings are required to obtain a d.c. voltage proportional to velocity. The device is very convenient and widely used in control systems where velocity feedback is required, but is unsatisfactory for precision measurement because of unavoidable ripple (due to the finite number of poles) and because of spikes due to the slip rings. The a.c. tacho is better in this respect. It consists of a rotating cylinder and two coils at right angles, one excited at constant frequency and voltage; the second coil detects a voltage proportional to the rate of rotation of the cylinder, owing to eddy current effects. The two transducers are illustrated in figure 5.21(a) and (b).

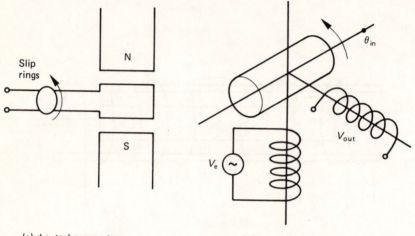

(a) d.c. tachogenerator   (b) a.c. tachogenerator

Figure 5.21 *A.c. and d.c. tachogenerators*

5.4 Strain transducers

We will discuss piezo-electric transducers and strain gauges in this section. Both these devices are often used for the measurement of force or acceleration, but the actual physical mechanisms involve a change in length so that it is appropriate to discuss them here.

5.4.1 *Piezo-electric transducers*

Piezo-electric transducers have been with us for a long time, and while this characteristic is usually derogatory (especially of female politicians and American presidents) piezo-devices have increased in popularity in recent years. As explained above, the basic physical effect is the production of a surface charge density in response to deformation; the magnitude depends on the direction with respect to the crystal axes and is specified by the d and g coefficients (section 3.2.2).

When considered as a length transducer, the physical process is summarised by the equation

$$Q = K \, \delta t \tag{5.2}$$

where Q is the charge, δt the deformation and K a constant related to the d and g coefficients. Considering a small disc of piezo-electric material of capacitance C, as in figure 5.22, the bulk modulus E of the material is given by

$$E = \frac{F/A}{\delta t/t}$$

so

$$K = \frac{Q}{\delta t} = \frac{dF}{\delta t} = \frac{dEA}{t} = \frac{dEC}{\epsilon \epsilon_0} = \frac{EC}{g}$$

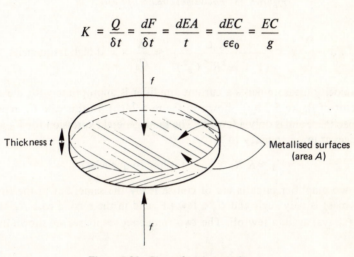

Figure 5.22 *Piezo-electric transducer*

The equivalent circuit of figure 5.23 (a) can be deduced from equation (5.2). It is more useful to choose a current generator i proportional to velocity than a charge generator proportional to displacement; the capacitance C appears in parallel with a large leakage resistance which can usually be ignored. The Thévenin equivalent in figure 5.23(b) is also useful, placing C in series with a voltage generator proportional to displacement. For a disc of quartz of area 1 cm^2 and thickness 1 mm^2, $C \approx 4$ pF, $K \approx 2 \times 10^{-2}$ A/ms^{-1} and $K/C \approx 5 \times 10^9$ V/m, which are very high in comparison with the responsivities found above for displacement transducers, though the applications are not really comparable.

The device can be operated so as to give essentially a response flat to either displacement or velocity, by choice of an appropriate amplifier. For displacement measurements a charge amplifier is used, as shown in figure 5.24(a) with the equivalent circuit of figure 5.23(b). A large resistance R_a is used for bias as with the capacitive displacement transducers. The output V_0 is given by

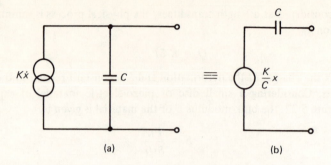

Figure 5.23 *Equivalent circuits of figure 5.22*

$$V_0 = \frac{K}{C}x \times \frac{\omega R_a C}{(1 + \omega^2 R_a^2 C_a^2)^{\frac{1}{2}}} \approx \frac{K}{C_a} \times x \text{ at high frequencies}$$

For velocity measurements a current amplifier is appropriate with the equivalent circuit of figure 5.23(a), shown in figure 5.24(b). In this case C_b represents stray capacitance and is only a few pF. The same general expression for V_0 applies, but reduces at low frequency to

$$V_0 \approx K R_b \dot{x} \qquad \text{(at low frequencies)}$$

The two amplifier circuits are of course really the same, but in the first case R_a (for bias) is very high and C_a a few pF, and in the second case R_b is lower and C_b (strays) again a few pF. The two frequency responses are shown in figure 5.25.

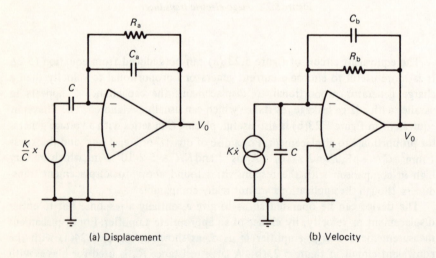

(a) Displacement   (b) Velocity

Figure 5.24 *Amplifier circuits for piezo-electric transducers*

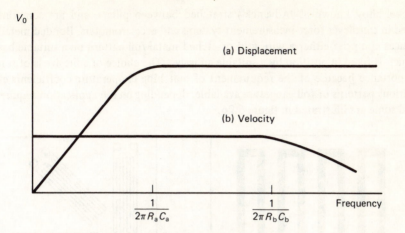

Figure 5.25   *Response of circuits of figure 5.24*

A useful velocity response over a wide frequency range (including d.c.) is thus possible and the device finds wide application in miniature velocity meters. However, it is important to note that the displacement response falls to zero for d.c., and although widely used in accelerometers the response usually falls off below a few tens of Hz.

5.4.2 *Strain gauges*

Strain gauges are essentially thin strips of material whose resistance changes when strained. We saw in section 3.3.2 that the basic equation of these transducers is given in terms of the gauge factor GF (ratio of fractional change in resistance to strain) as

$$GF = \frac{\delta R/R}{\delta l/l} = 1 + 2\,\nu + \frac{\delta \rho/\rho}{\delta l/l}$$

where ν is Poisson's ration and ρ the resistivity. The second term is known as the dimensional term, and the third as the piezoresistive term.

Strain gauges fall naturally into two types — metallic and semiconductor. For metals ν is approximately one-half and the piezoresistive effect is small, so $GR \approx 2$, whereas for semiconductors (usually based on silicon) the piezoresistive effect is large (≈ 100) and totally dominant. In fact $\delta \rho/\rho$ is proportional to stress, so GF is proportional to the bulk modulus E. Unfortunately the high value of GF is accompanied by a high temperature coefficient which limits the application of semiconductor devices.

Metallic gauges may be bonded or unbonded. The latter are simply thin wires ($\approx 25\ \mu m$) of a material of low temperature coefficient (for example, a copper-nickel alloy known as 'Advance') stretched between pillars, and are sometimes used in cantilever force-measurement systems and accelerometers. Bonded metallic gauges comprise either a wire grid or etched metal foil pattern on a suitable base, which is fixed in position by a suitable adhesive. The choice of adhesive is of some importance because of the requirement of matching temperature coefficients etc. Various patterns of foil gauges are available, depending on the application required, and some are illustrated in figure 5.26.

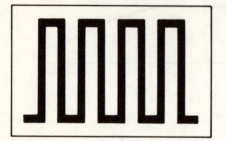

(a) Linear

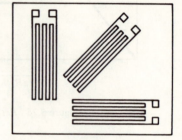

(b) Rosette

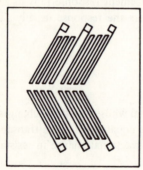

(c) Torque

(d) Diaphragm

Figure 5.26 *Strain gauge transducers*

The main source of error in measurements with strain gauges is due to temperature changes, especially with semiconductor devices. Several gauges are used together, some of them unstrained, usually in a four-arm bridge configuration. Two such arrangements are shown in figure 5.27.

A typical amplifier configuration is shown in figure 5.28. The four gauges G_1, G_2, G_3 and G_4 are nominally identical and of unstrained resistance R and connected differentially, so that the equivalent circuit is as shown. The output

$$V_0 = V_e \frac{\delta R}{R} \times \frac{R_f}{R/2} = V_s \times (\text{strain}) \times \frac{R_f}{R/2} \times \text{GF}$$

is directly proportional to strain. The maximum permissible strain is usually about half a percent, and it is usual to arrange that the amplifier limits before this value is reached.

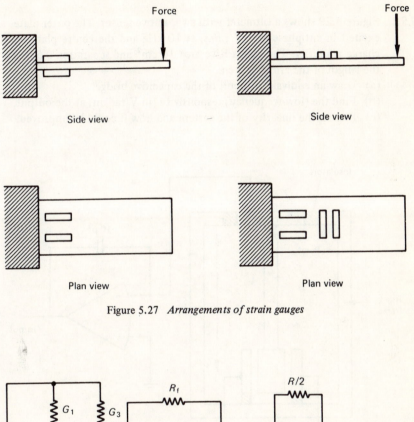

Figure 5.27 *Arrangements of strain gauges*

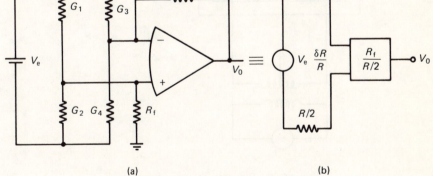

Figure 5.28 *Strain gauge amplifier (a) and equivalent circuit (b)*

5.5 Exercises on chapter 5

5.5.1. Discuss the relative advantages and disadvantages of LVDTs and variable-area capacitive transducers for the measurement of displacements of less than 1 μm in ranges up to 1 cm.

5.5.2. Figure 5.29 shows a tiltmeter with a capacitive sensor. The outer plates are excited in antiphase at 1 V r.m.s. at 10 kHz and the centre plate feeds a charge amplifier. The plates have area 10 cm^2 and separation 0.5 mm and the length of the arm L is 5 cm.
(a) Draw an equivalent circuit of the capacitive bridge.
(b) Find the (low-frequency) responsivity (in V/radian) at the output.
(c) Discuss the linearity of the system and how it could be improved.

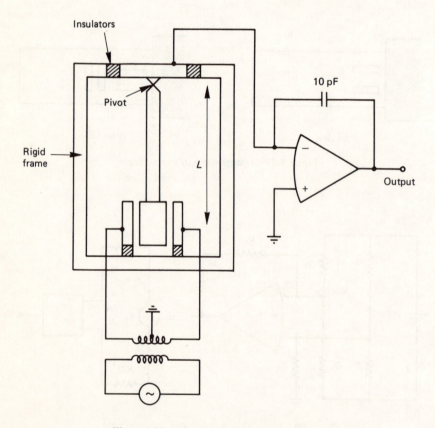

Figure 5.29 *Tiltmeter with capacitive sensor*

5.5.3. Figure 5.30 is a schematic diagram of a digital micrometer, which is to have a range from 0 to 30 mm and a detectivity of 0.01 mm. Discuss the feasibility of using Moiré gratings as the basic sensing element, pointing out possible problems in the mechanical design.

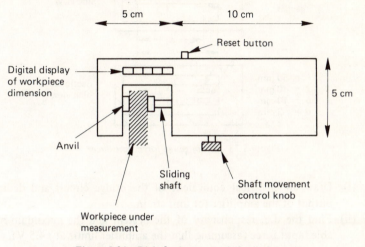

Figure 5.30 *Digital micrometer (schematic)*

5.5.4. A digital readout is required from a wind speed and direction system, comprising 'weather-vane' and rotating-cup devices, by attaching suitable transducers. Discuss the most appropriate transducers, making reasonable assumptions about the accuracy and resolution required.

5.5.5. Illustrate the terms *responsivity*, *detectivity*, *range* and *accuracy* by reference to the applications below, discussing the type of transducer required in each case.
  (i) The displacement of the suspended mass in a seismometer is to be measured with an overall responsivity of 10^7 V/m. The maximum displacement is ± 10 μm and changes of $\pm 10^{-12}$ m in a bandwidth of 1 Hz must be detectable.
  (ii) It is required to detect displacements of ± 10 μm in a shaft which moves linearly over a distance of ± 1 cm with maximum velocity 1 cm/s. An overall responsivity of 100 V/m is required.

5.5.6. Figure 5.31 shows a schematic strain gauge force transducer, in which a force f is applied to the end of a cantilever beam of mild steel (modulus 2×10^{11} N/m^2). Four metallic strain gauges (gauge factor 2) are attached to the upper and lower faces as shown and are connected differentially in a bridge circuit excited at 9 V d.c. and feeding an amplifier of gain 200.

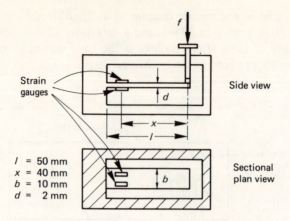

Figure 5.31 *Strain gauge force transducer*

(i) Draw an electrical equivalent of the bridge circuit and deduce the output of the amplifier for unit strain.

(ii) Find the d.c. responsivity of the system and the maximum measurable input force (assuming that the amplifier limits at ±4.5 V), and the strain for this force.

Note: A force f at the free end produces surface stress at a distance x from the free end of $6fx/(bd^2)$.

6

Basic Transducers for Temperature

6.1 Scale of temperature

Although temperature is very apparent to our senses, it is far less easy to define a sensible scale for temperature than for, say, length. This is because we detect or measure temperature by the effect it produces, such as the expansion of a liquid, so an assumption has to be made about the linearity of the effect observed. The temperature of a system is a measure of its energy; this is best described by the Principle of Equipartition, which states that for a system in thermal equilibrium with its surroundings the mean kinetic energy per degree of freedom is $\frac{1}{2}kT$, where k is Boltzmann's constant and T is the absolute temperature.

The most direct realisation of a scale of temperature using this idea is based on the gas equation for an ideal gas, which can be described directly from the Principle of Equipartition by means of the Kinetic Theory of Gases. The gas equation $pv = RT$ gives the pressure p of a given mass of gas of volume v at temperature T, with a constant of proportionality R which is equal to the number of molecules multiplied by k. The constant volume gas thermometer, in which the pressure of a fixed volume of an inert gas is measured as a function of temperature, is the basis of the thermodynamic scale of temperature. This is the nearest one can get to an absolute scale of temperature based on the idea of temperature as a measure of energy. On this scale the only fixed point is the triple point of water, defined as 273.16 degrees Kelvin and as 0.01 degrees Celsius ('by international agreement' is the excuse for the silly number).

Absolute measurements with gas thermometers are made at a few specialised laboratories, but for practical purposes a set of fixed points has been designed, following accurate measurements with gas thermometers, and forms the International Practical Scale of Temperature. The idea is that the various fixed points can be reproduced relatively easily and interpolation between them made by specified transducers. Table 6.1 shows some of the fixed points.

81

Table 6.1   Fixed points on the International Practical Scale of Temperature

Temperature ($^{\circ}C$)		
−182.97	BP of oxygen	⎫
0.01	Triple point of water	⎬ Platinum resistance thermometer
100.0	BP of water	⎪
444.6	BP of sulphur	⎭
960.8	FP of silver	⎫ Platinum/platinum–rhodium
1063.0	FP of gold	⎬ thermocouple

6.2   Temperature transducers

There are not many types of temperature transducer. The two main classes are resistive devices, including both metallic and semiconductor types, and thermo-electric devices. Resistive transducers are modulating, and require a bridge circuit to produce a usable output, whereas thermo-electric transducers are self-generating.

6.2.1   Resistive temperature transducers

These transducers are rather like wire-wound resistors, being in the form of a non-inductively wound coil of a suitable metal wire, usually platinum, copper or nickel. They are often encapsulated in a glass rod in the form of a probe. The nominal resistance may vary between a few ohms and a few kilohms, but 100Ω is a fairly standard value. As seen in chapter 3, the relationship between resistance and temperature is nearly linear, but is more accurately described by equation (6.1)

$$R_T = R_{T_0}(1 + aT + \beta T^2 + \gamma T^3) \tag{6.1}$$

a is typically about 0.4×10^{-2} for platinum, but β and γ are very small (of the order 10^{-6} and 10^{-12} respectively). The relationship is shown in figure 6.3, where it is compared with that for thermistors.

The devices are useful over a temperature range from about -200°C to about 300°C for nickel and copper, and up to 900°C for platinum. Their main advantages are that they are stable and reproducible with high linearity, but they are relatively large, so have a long time constant. Their responsivity is low, ranging from $0.4\Omega/^{\circ}$C for platinum to about $0.7\Omega/^{\circ}$C for nickel (for a 100Ω device).

In the last few years improved devices have become available, in which a rectangular matrix of platinum is deposited on a ceramic substrate and connections in the matrix are trimmed by laser to produce a resistance of precisely 100Ω at 300°C. A typical transducer measures 3 cm $\times$ 5 mm $\times$ 1 mm, and a schematic is shown in figure 6.1.

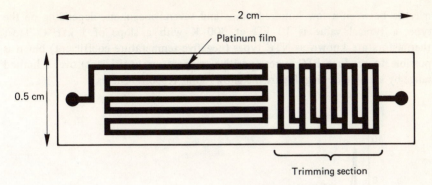

Figure 6.1 *Miniature platinum resistance element*

The transducers are operated with conventional bridge circuits, as shown in figure 6.2, but compensating leads are necessary because of the low element resistance.

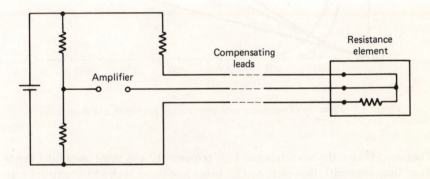

Figure 6.2 *Bridge circuit for resistance thermometer*

6.2.2   Thermistors

Thermistors are small semiconducting transducers, usually in the shape of beads, discs or rods. They are made by sintering mixtures of oxides of various materials such as cobalt, nickel and manganese and are often encapsulated in glass. As we saw in chapter 3 they are inherently non-linear, the resistance following an equation of the form

$$R_{T_1} = R_{T_0} \exp \left[\beta \left(\frac{1}{T_1} - \frac{1}{T_0} \right) \right]$$

where β is a constant of value about 3000. The resistance at room temperature may be between a few hundred ohms and several megohms, depending on the type; a typical value is 10 kΩ at 300 K with a slope of 1 kΩ/°C. Most thermistors are known as NTC types (negative temperature coefficient) but it is possible to produce PTC types (positive temperature coefficient) over a limited range by suitable doping. Figure 6.3 shows some response curves.

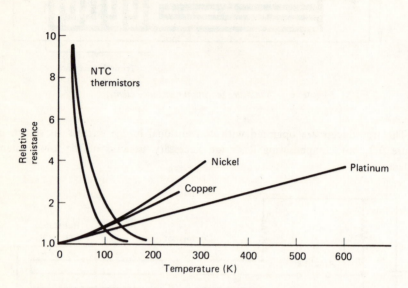

Figure 6.3   *Variation of resistance with temperature for metals and thermistors*

Thermistors have the advantages of high responsivity and small mass (and hence short time constant). However, besides being non-linear their characteristics may vary appreciably from sample to sample and they have a limited temperature range of about −100°C to 200°C. Non-linearity is no longer considered to be much of a disadvantage, since transducers are being more often connected directly to microprocessors, so that the response characteristics can be compensated in software. It is possible to linearise the response by electronic means, which essentially involves placing a suitable resistor in series, and an overall linearity of about 5 per cent over a range between 0°C and 100°C can be obtained. However, the most suitable application for thermistors is in temperature control, where their linearity is less important and where their high responsivity enables very high detectivities to be achieved.

Thermistors are usually operated in bridge circuits, and compensating leads are not usually needed because of their reasonably large resistance. Some care has to be taken in limiting the power dissipated, which can seriously affect the measurement because of their small thermal capacitance. Figure 6.4 shows a typical bridge arrangement.

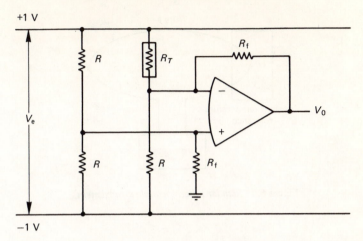

Figure 6.4 *Thermistor bridge circuit*

It is easy to analyse the above circuit, by means of the equivalent circuit discussed in chapter 4.

$$V_{\text{out}} = \frac{V_{\text{ex}}}{4} \times \frac{\delta R}{R} \times \frac{R_{\text{f}}}{R/2}$$

and with $V_{\text{ex}} = 2$ V, $R = 10$ kΩ, $\delta R = 1$ kΩ and $R_{\text{f}} = 100$ kΩ, we have a responsivity of 1 V/°C. Changes in temperature as small as 0.01°C can be detected very easily, and it is possible with a similar circuit to obtain a detectivity as high as 10^5/°C.

6.2.3 Thermo-electric transducers

Thermo-electric transducers are known as thermocouples, and are self-generating transducers comprising junctions between two dissimilar wires as described in chapter 3. The conventional arrangement is shown in figure 6.5, where one junction is maintained at 0°C (in melting ice) and the other attached to the object to be measured.

This arrangement is rather inconvenient, because of the layout of the leads and the melting ice container. A more convenient practical arrangement is shown in figure 6.6. The two wires are laid out side by side and are connected directly to the voltmeter. The two junctions between the wires and the terminals of the voltmeter do not produce any error if they are both at the same temperature, but there is no proper reference junction. These two junctions are in fact the reference, so an error tends to occur if the temperature of the room changes. This is avoided by using cold-junction compensation, in which the gain of the amplifier is modified

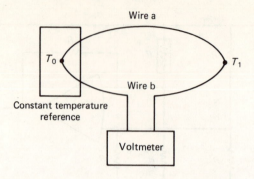

Figure 6.5 *Standard thermocouple arrangement*

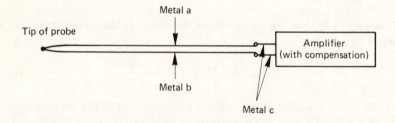

Figure 6.6 *Thermocouple arrangement with cold-junction compensation*

so as to cancel the error. A resistance temperature device is included in the amplifier circuitry and changes the gain appropriately.

The arrangement in figure 6.6 is very convenient and almost always used in practice. A flexible stainless steel sheath often encloses the wires to improve the robustness, though at the cost of an increased time constant.

The most popular materials are platinum/platinum–rhodium, which is the standard device, copper/Constantan and Chromel/Alumel. The output is basically very low, ranging from about 10 μV/°C for the former to 80 μV/°C for the latter. Indeed the fact that the transducer is self-generating is not very significant because an amplifier is always needed in practice. The low output means that it is difficult to detect small changes in temperature and a detectivity of better than 10/°C is rare.

The main advantage of thermo-electric transducers is their wide temperature range, nominally −184°C to 1260°C for Chromel/Alumel and up to 2000°C or more for some alloys of tungsten.

6.2.4 Solid-state devices

p–n junctions have become popular temperature transducers in the last few years. They rely on the forward-voltage/temperature coefficient of a silicon diode, which is about -2 mV/°C. Unfortunately the exact value of the effect varies considerably between samples, so calibration is necessary. However, the effect is very linear and the transducers are extremely cheap.

A typical circuit is shown in figure 6.7. The diodes are usually transistors with base and collector connected together. Strictly they should be operated under constant current, but in practice the error is small.

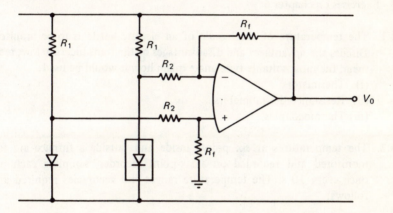

Figure 6.7 p–n *junction temperature transducer*

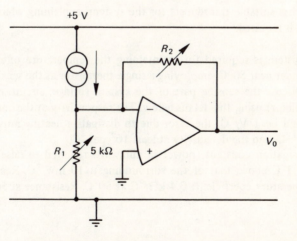

Figure 6.8 *'Current source' temperature transducer*

The same effect is used in miniature encapsulated solid-state devices which are now available. They are sometimes called 'current sources' and are small integrated-circuit two-terminal devices usually calibrated to produce a current of 1 $\mu A/K$, so they produce 300 μA at 300 K. They have a working range of about $-50°C$ to $150°C$, over which the linearity is about 1 per cent, but unfortunately are rather expensive. Figure 6.8 shows a device arranged to give a responsivity of 0.1 V/°C over the range 0°C to 100°C (by trimming the two resistors R_1 and R_2 respectively).

6.3   Exercises on chapter 6

6.3.1. The temperature of the handle of an electric kettle is to be monitored. Discuss the advantages and disadvantages of the transducers below, recommend the most suitable type and explain how it would be used.
(i)   Thermistors
(ii) Resistance thermometers
(iii) Thermocouples.

6.3.2. The temperatures at six points inside and outside a furnace are to be monitored and recorded on a six-point recorder, sampling each point once every 10 s. The temperature ranges and accuracies required are as follows:

Internal: 3 points in range 500–1000°C, accuracy ±5°C
External: 1 points in range 50–100°C, accuracy ±0.5°C
            2 points in range 25–35°C, accuracy ±0.1°C

Discuss suitable transducers for the system, explaining what instrumentation is required.

6.3.3. A system is required for maintaining the temperature of a small copper cylinder near 50°C, employing a single thermistor as the sensing element.
    Discuss the sensing part of the system (bridge, circuit, excitation and amplifier) using the details below. The responsivity at the amplifier output should be 1 V/°C, the error due to dissipative heating must be less than $10^{-2}°C$, and the detectivity at least 10^2 per °C.
[Dissipation constant (power in milliwatts required to raise the temperature 1°C above that of the surroundings): 10 mW/°C. Resistance versus temperature coefficient: 0.4 kΩ/°C at 50°C. Resistance at 50°C: 10 kΩ.]

7

Basic Tranducers for Radiation

7.1 Classification of photodetectors

The term photodetectors is usually applied to transducers for a very narrow range — the visible and near infra-red — of the electromagnetic spectrum. There is of course no real limit to the electromagnetic spectrum; it stretches from very low frequency waves of less than 1 Hz, observed by magnetometers, to X-rays and γ-rays of frequencies up to 10^{20} Hz, as indicated in figure 7.1. However, we will confine our attention here to photodetectors, since this range is particularly important in many types of measurement.

The standard of light is the *candela*, defined as one-sixtieth of the luminous intensity of a full radiator, which is a 'black body' at a temperature of 2042 K (when it looks anything but black). This is not particularly useful for measurement purposes since one is usually more interested in absolute values, measured in Watts, than in the effect on the eye to which the formal definition is related (a luminous intensity of one candela is in fact equivalent to 0.00146 W per steradian). We will always use Watts when discussing responsivity and detectivity.

There are only two main types of photodetector, namely thermal detectors and photon detectors. The former consist essentially of a 'black' screen which ideally absorbs all the radiation incident on it and therefore rises in temperature. The temperature rise is measured by some form of transducer; the photodetectors are usually named after the particular type used, thermocouple and pyro-electric devices being the most popular. Thermal photodetectors usually have a fairly low responsivity but it is essentially constant over a wide range of wavelengths (since all radiation is absorbed, independently of wavelength). This is in sharp contrast to photon photodetectors. These devices employ some form of photo-electric effect, such as the photo-emissive effect, the photoconductive effect or the photovoltaic effect, and their responsivity increases linearly with wavelength up to a maximum, beyond which it falls rapidly to zero. This is because the effect

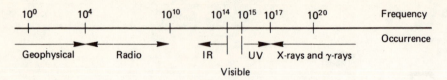

Figure 7.1   *The electromagnetic spectrum*

occurs only if the photon energy $h\nu$ of the incident radiation is greater than or equal to that of the relevant photo-effect, such as work function or energy gap, so the process is inefficient at short wavelengths (since only one photo-electron is produced per photon) and falls to zero beyond a critical wavelength.

7.2   Thermal photodetectors

As we saw in chapter 3, a stream of radiation incident on a collecting screen is analogous to the charging of a capacitor by a steady current, and the temperature rise due to an incident stream W Watts at frequency f Hz is given by

$$\delta T = \frac{\epsilon W / G}{(1 + 4\pi^2 f^2 \tau^2)^{\frac{1}{2}}} \qquad (7.1)$$

ϵ is the emissivity of the screen, G its thermal conductance and τ the time constant (C/G, where C is the thermal capacitance). We can use this equation to deduce the responsivities of the various thermal detectors.

7.2.1   *Thermocouple detectors*

Figure 7.2 is a schematic drawing of a thermocouple detector. It is essentially a thermocouple in which the collecting screen is an extended hot junction. The screen is usually made from thin blackened gold foil supported by an insulated ring, and the two contacts are of materials of thermo-electric power of opposite polarity to produce a larger responsivity.

The voltage produced for a temperature rise δT is $P\delta T$, where P is the difference in thermo-electric power, so from equation (7.1) the overall responsivity is

$$r = \frac{\epsilon P / G}{(1 + 4\pi^2 f^2 \tau^2)^{\frac{1}{2}}}$$

It can be seen that for large r we require G to be small, and this can be achieved by placing the detector in an evacuated jacket with a window of suitable optical

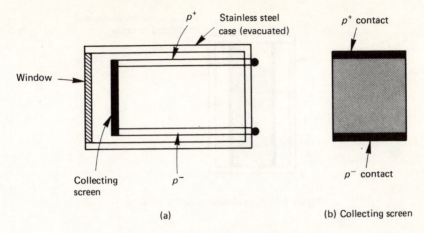

Figure 7.2 *Thermocouple radiation detector*

properties. Unfortunately this makes τ long, so a poor frequency response is obtained. Sometimes the jacket is filled with an inert gas to decrease τ, but at the cost of a lower responsivity.

Most thermocouple detectors have a small active area of 1–10 mm^2, a time constant of at least 10 ms and a fairly low responsivity of 10–100 V/W. The element resistance is typically 100Ω, making it difficult to obtain a good noise figure with the following amplifier. Their low responsivity and poor frequency response (up to about 10 Hz) limit their application mostly to calibration instruments such as spectrometers where absolute measurements of incident radiation are required.

7.2.2 *Pyro-electric detectors*

These devices form a useful complement with thermocouple detectors, since they operate only at frequencies above about 10 Hz. They consist of a thin slice of pyro-electric material, usually lead zirconate titanate, with metallised surfaces and a protective window, as shown in figure 7.3.

The rise in temperature δT is transduced to a surface change density $Q = K\delta T$ where K is a constant. It is more convenient to work in terms of current, which is given by

$$I = \frac{2\pi f \times K \times \epsilon W/G}{(1 + 4\pi^2 f^2 \tau^2)^{\frac{1}{2}}}$$

The responsivity r can be written

$$r = \frac{\epsilon K}{C} \times \frac{2\pi f \tau}{(1 + 4\pi^2 f^2 \tau^2)^{\frac{1}{2}}}$$

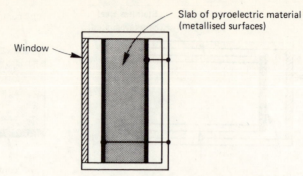

Figure 7.3 *Pyro-electric radiation detector*

This tends to zero at low frequencies but becomes constant and of value $\epsilon K/C$ for $4\pi^2 f^2 \tau^2 \gg 1$. An equivalent circuit is shown in figure 7.4, comprising a current generator rW in parallel with the *electrical* capacitance C_1.

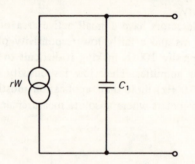

Figure 7.4 *Equivalent circuit of pyro-electric detector*

Since the transducer is capacitive it normally feeds a charge amplifier which is often integral with the detector since its noise level determines the overall detectivity (because pure capacitors do not produce electrical noise).

Typical devices have active areas between 1 and 10 mm^2, thermal time constants τ about 100 ms, electrical capacitance 10–100 pF and responsivity about 10^{-4} A/W. The frequency response is flat from about 10 Hz to a few kHz. The windows used limit the spectral response to about 1–10 μm.

7.2.3 Other thermal detectors

The two devices described above are the most widely used, but a few other types have been developed. The Golay cell was developed many years ago and employs the unlikely sounding principle of expansion of a small flexible cell of inert gas. It is comparable in characteristics to the thermocouple detector. The thermistor bolometer is used in some applications, and uses a thermistor to detect the temperature rise of the collecting screen. A superconducting version has an exceptionally high responsivity and detectivity, though it is used only in special circumstances of course (for example, if the establishment has a lot of money).

7.3 Photon detectors

7.3.1 Photo-emissive detectors

Photo-emissive detectors are rather like radio valves in appearance, which the writer understands to have comprised an anode and cathode inside an evacuated jacket and to have been used in early radio sets before the advent of transistors and integrated circuits. In figure 7.5 the cathode is composed of a suitable metal

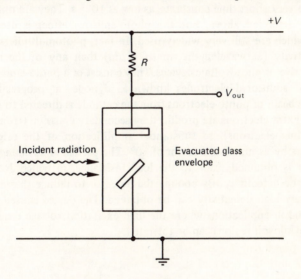

Figure 7.5 *Photo-emissive detector*

which emits electrons when radiation W is incident, provided that the photon energy $h\nu$ is greater than the work function ϕ, and these are drawn off to the anode, producing a photocurrent I given by

$$I = \frac{eqW}{h\nu}$$

where e is the electronic change and q an efficiency factor, known as the quantum efficiency. The current flows through a load resistance R, producing an output voltage which may be amplified as required. The responsivity can be deduced in terms of current or voltage, and is

$$\text{(current)} \qquad r_i = \frac{eq}{h\nu} \qquad \text{A/W}$$

$$\text{(voltage)} \qquad r_v = \frac{eqR}{h\nu} \qquad \text{V/W}$$

Practical quantum efficiencies are rather low, often as small as 0.01, but the responsivities obtained are nevertheless fairly high. With a load resistance of 1 MΩ, r_v is about 10^4 V/W at 1 μm. In fact 1 μm is about the largest usable wavelength. Very low work functions cannot be obtained, and a combination of caesium oxide and silver is the best-known cathode material.

Photo-emissive detectors have the rather serious disadvantage of being rather bulky and fragile, and of requiring fairly high voltages of 100 V or more, though they do have very short time constants, as low as 10^{-8} s. They are not much used in the form described above, but the photo-emissive effect is used in photo-multipliers which are still very widely used. In fact, photomultipliers still have a better detectivity (at wavelengths up to 1 μm) than any of the more recent photoconductive or photovoltaic devices. They consist of a photo-emissive cathode and a set of additional electrodes known as dynodes at progressively higher voltages. The beam of photo-electrons from the cathode is directed to the dynodes in turn, and extra electrons are produced by secondary emission (from the energy of the colliding electrons), so substantial amplification of the original photo-current occurs by as much as a factor of 10^6. The particular advantage is that the amplification is obtained prior to the load resistance, in which Johnson noise usually sets the detectivity. By cooling the detector to reduce thermally excited emission a very high detectivity can be obtained. The device is used in very low light levels and in applications where the high excitation voltage required (which may be in the kilovolt region) can be tolerated.

7.3.2 Photoconductive detectors

Photoconductive detectors are composed of semiconductor materials in which incident photons cause excitation of electrons across the energy gap, leading to a change in conductivity. It is important to note that the excited electrons have a finite lifetime so that when illuminated by a steady beam an equilibrium condition is rapidly reached in which the decay of the increased number of electrons in the conduction band just matches the excitation. The essential action is thus an increase in conductivity over that in the absence of illumination, and it has been shown previously that the fractional change in bulk resistance is given by

$$\frac{\delta R}{R} = \frac{qW\tau}{h\nu N}$$

where q is the quantum efficiency, W the incident radiant power at spectral frequency ν, τ is the lifetime, h is Planck's constant and N the number of electrons in the conduction band in the absence of radiation. N is governed by thermal excitation, being given by

$$N = N_0 \, \exp\left(-E_g/2kT\right)$$

where N_0 is the total number of electrons and E_g the energy gap. N therefore rises with temperature and will be relatively large at a given temperature if E_g is small.

Photoconductive detectors are modulating transducers and are operated in a bridge circuit. Figure 7.6 shows a typical arrangement and its equivalent circuit, and it can be seen that the responsivity is given by

$$r = \frac{V_{ex}}{4} \times \frac{q\tau}{h\nu N} \tag{7.2}$$

(the condition load resistance R = cell resistance R_c gives the optimum responsivity).

The responsivity therefore rises linearly with increasing wavelength up to a maximum when $h\nu = E_g$ (that is, $\lambda = hc/E_g$ where c is the velocity of light) and then falls rapidly to zero. Unfortunately equation (7.2) contains several terms whose exact values are not easy to determine, such as q, τ and N, so that although we can deduce the characteristics of these detectors we cannot find their exact responsivities theoretically. To obtain a high responsivity we clearly require a long lifetime τ, high efficiency q and a small number of electrons N, equivalent to a high resistance R_c in the absence of radiation. If we want a response far into the infra-red the energy gap E_g must be small, so at room temperature N will be high and R_c small. Apparently we cannot have both high responsivity and good long-wavelength response together. Rather surprisingly, since munificence is not often exhibited in physics (and even more rarely in physicists)*, it is possible to obtain both together at the complication of cooling the detector, usually to liquid nitrogen temperature (77 K). This greatly reduces the thermal excitation, increasing the cell resistance considerably and hence the responsivity.

Figure 7.7 shows the variation of responsivity with wavelength for some popular devices, together with their energy gaps and corresponding critical wavelengths. The specific values of responsivity depend on the design of the particular specimen used (area, shape etc.) since the cell resistance is a controlling factor, so the values shown in the curves are for comparison purposes.

* The author is of course writing in the role of an engineer in this chapter.

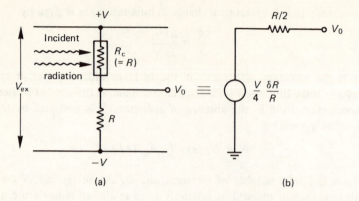

Figure 7.6 *Bridge circuit for photoconductive detector (a) and equivalent circuit (b)*

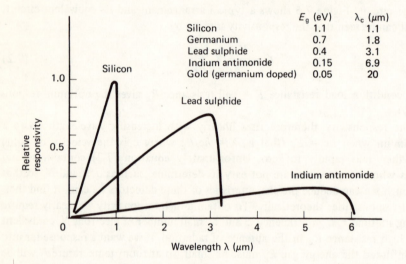

	E_g (eV)	λ_c (μm)
Silicon	1.1	1.1
Germanium	0.7	1.8
Lead sulphide	0.4	3.1
Indium antimonide	0.15	6.9
Gold (germanium doped)	0.05	20

Figure 7.7 *Variation of responsivity with wavelength (room temperature)*

Lead sulphide is one of the best-known general-purpose detectors. It is available in areas of about 1 mm² to a few cm², usually in the shape of rectangular or circular deposits on a glass of ceramic substrate, as shown in figure 7.8(a). Cell resistances may be as high as 1 MΩ and time constants fairly long ($\approx$ 100 μs), so responsivities can be 10^4 to 10^5 V/W but with rather poor frequency response (up to a few kHz). Cooling is not usually necessary.

Indium antimonide is used for more specialised applications. It has the smallest energy gap, and therefore the best infra-red response, of any undoped material, and also has a very short time constant. It is difficult to use at room temperature, having a resistance of only a few tens of ohms, though such devices are available.

However, when cooled it has a responsivity of about 100 V/W and is widely used in remote sensing applications (for example, in thermal scanning of the earth's surface from aircraft and spacecraft) where its good frequency response and excellent spectral response make it ideal. Cooled detectors are similar in appearance to lead sulphide detectors (except that you cannot actually see them because of the attached cooling Dewar). Uncooled detectors are often in the form of a long thin matrix pattern to increase the resistance, as shown in figure 7.8(b).

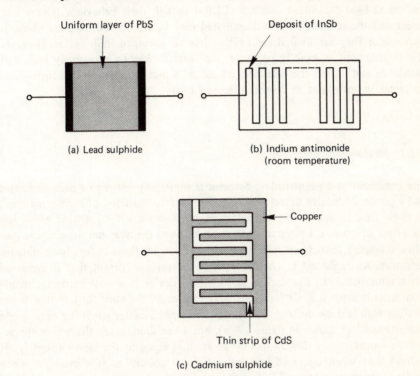

Figure 7.8 *Various photoconductive detectors*

The effect described above is strictly the *intrinsic* photoconductive effect. *Extrinsic* detectors can be produced by doping, which can place extra energy levels in the otherwise forbidden energy gap, so photons of lower wavelength may produce excitation. Efficiencies tend to be very low and cooling is essential, but it is possible to obtain a response far into the infra-red. Germanium doped with gold responds as far as 20 μm, as shown in the table in figure 7.7.

Another important class of photoconductive detectors employ the charge-amplification effect. As explained in chapter 3, impurities can cause hole-trapping effects and result in a very long effective lifetime, though with a similarly increased time constant of course. The two best known materials are cadmium sulphide (CdS) and cadmium selenide (CdSe). Both have large energy gaps of about 2 eV, giving critical wavelengths at 0.6–0.7 μm; the response of CdSe matches that of the eye quite well. Detectors using either of these materials are characterised by very high cell resistance (several MΩ) and responsivity (up to 10^7 V/W), and long time constants (≈ 0.1 s). The reproducibility between devices is very poor, since the effect depends on impurities, and they are mostly used in switching applications rather than for accurate measurement. They are often known as light-dependent resistors (LDRs) though their behaviour is very non-linear and the resistance when illuminated may be only a few hundred ohms. In appearance they are at first sight rather like an uncooled InSb device. However, the requirement is exactly opposite; the material has an excessively high dark resistance and in fact the detectors consist of a thin strip of material surrounded by metal on both sides, as shown in figure 7.8(c).

7.3.3 Photovoltaic detectors

The behaviour of a photovoltaic detector is something between a photo-emissive and a photoconductive detector. The devices are constructed of similar materials to those used in photoconductive devices, but contain a p–n junction which has the effect of causing a physical separation between the hole and the electron pair when subjected to incident radiation, so that a current flows as for photo-emissive detectors. As explained in chapter 3, it is the reverse current that is increased when illuminated. In the dark a photovoltaic device is almost indistinguishable from an ordinary p–n diode (and its characteristics are the same too), though it has a rather high leakage current. The current/voltage characteristic is the usual diode characteristic, as shown in figure 7.9(a), but when illuminated the whole charac-teristic moves bodily downwards by an amount equal to the light current I_L. It follows that when operated under short-circuit conditions (for example, when feeding a current-measuring device) the light current I_L is observed, whereas when operated open-circuit (for example, when feeding a voltage-measuring device) a voltage V_{oc} is seen. The transducer is self-generating, since it produces a voltage/current even in the absence of an external power supply or bias, when it is of course the well-known solar cell used for power generation in spacecraft etc. However, it is usually operated as a modulating transducer at a small reverse bias since the characteristic is more uniform in that region. Also, the capacitance is reduced by reverse bias, improving the frequency response.

A shown in chapter 3, the light current $I_L = eqW/h\nu$, so the responsivity is

$$r = \frac{eq}{h\nu} \quad \text{A/W}$$

The efficiencies can be quite high, up to 30 per cent, and the responsivity is typically 0.1 A/W at 1 μm. A useful equivalent circuit is shown in figure 7.9(b), in

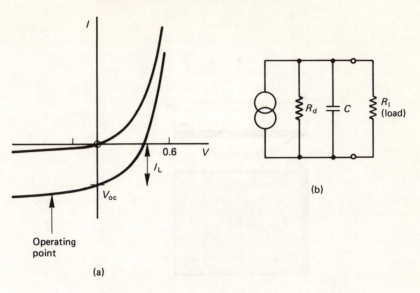

Figure 7.9 *Current/voltage characteristics of photodiode (a) and equivalent circuit (b)*

which the light current is divided between the diode resistance R_d and the load resistor R_1.

The transducer is produced by first depositing a thin layer of, say, *p*-doped material on a suitable substrate followed by a very thin layer of *n*-doped material, forming an extended junction, as shown in figure 7.10. The only common device is silicon-based, normally referred to as the silicon solar cell, and has a critical wavelength of about 1 μm. It is available in a wide range of sizes, from area of 1 mm² or less up to several cm². The junction capacitance is substantial, often between 100 pF and 1000 pF, and may limit the frequency response.

The transducer may be operated in voltage or current mode, but the latter is usually preferred since the light current is directly proportional to intensity, whereas the relationship between voltage and intensity is not linear. A simple circuit is shown in figure 7.11, though it is capable of excellent responsivity and detectivity. The output to steady illumination can be adjusted by means of the variable resistor. With $W = 10$ μW, responsivity 0.1 A/W at 1 μm and $R_f = 1$ MΩ, the voltage output is 1 V, giving an overall responsivity of 10^5 V/W.

Photovoltaic detectors are very widely used, being semiconductor devices directly compatible with most electronic circuitry. The detector is sometimes deposited on the same chip as an amplifier, providing a small compact sensor. A large number of devices based on the effect are available.

Phototransistors are similar to ordinary junction transistors but light is allowed to fall on the collector–base region, producing a base current which is amplified by the transistor. These devices are compact and useful in some applications, but

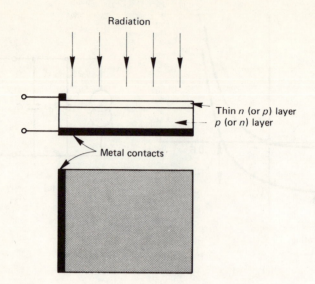

Figure 7.10 *Photodiode detector*

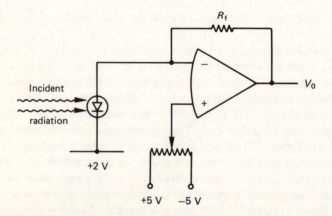

Figure 7.11 *Amplifier circuit for photodiode*

they have a poor frequency response and large dark current, and much better characteristics can be obtained by a separate *p–n* junction and amplifer.

Duodiodes are essentially phototransistors without a base connection. They are a *p–n–p* or an *n–p–n* sandwich and act rather like (non-linear) photoconductive devices. They have the interesting property that the polarity does not matter, which is very valuable when students have access to them. They are often of a cylindrical construction of small diameter with a lens in the end, and are used in card readers, bar-code readers etc.

Pin photodiodes are *p–n* devices with an extra layer of intrinsic (undoped) material between the *p* and *n* layers. This increases the junction area, so increasing the efficiency (and hence responsivity), decreasing the capacitance (and improving the frequency response) and reducing the reverse current (increasing the detectivity). They are very valuable in low light-level applications and where rapid response is necessary.

Photofets are FETs with light falling on the gate region, and are noted for their large gain–bandwidth product. Avalanche photodiodes employ a breakdown multiplication which produces high gain and low noise, rather like the effect in photomultipliers described above. However, these two devices tend to be expensive and are used only in specialised applications.

7.4 Exercises on chapter 7

7.4.1. An intensity monitor is required for measuring illumination levels in a factory. Discuss the advantages and disadvantages of the following transducers, recommend the most suitable type and explain how it would be used.
   (i)   Cadmium sulphide photoconductive cell
   (ii) Silicon solar cell
   (iii) Pyro-electric cell.

7.4.2. Discuss the types of photodetector that could be used for each of the applications below, recommending the most suitable type in each case.
   (i)   A lamp that switches on automatically when darkness falls (you may assume that it does not switch off automatically as soon as light is present!).
   (ii) An intruder-detection system, operating by collecting the radiation emitted by the intruder.
   (iii) A photographic exposure meter.
   (iv) Measurement of the stability of intensity of a continuous laser beam (mean power about 5 mW at 0.6 μm).
   (v)   The detection of very small amounts of light released in a chemical reaction, in otherwise total darkness.

7.4.3. A light-emitting diode is distant L metres from a photodiode, as shown in figure 7.12, and emits 10 μW of radiation at 1 μm, uniformly distributed over a cone of angle 20°. The photodiode has an effective area of 1 cm^2 and responsivity 0.1 A/W at 1 μm. Find the signal at the output of the amplifier.

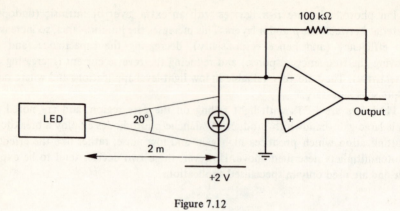

Figure 7.12

7.4.4. A small probe is required for detecting and locating overheated components in electronic circuits, by measuring the infra-red radiation emitted. The photodetector used should have an effective diameter of about 2 mm and be held about 3 cm from the component being checked.

Discuss the design of the system, making reasonable estimates of the parameters involved, and estimate the responsivity of your proposed system.

[The total radiation collected by a black body of area a_1 situated a distance d from a hot body of temperature T and area a_2 is $\sigma a_1 a_2 T^4 / \pi d^2$ where $\sigma = 5.6 \times 10^{-8}$ W/K^4 is the Stefan–Boltzmann constant. The peak emission from a body of temperature T occurs at about $2900/T$ μm.]

8

Other Transducers

In the previous three chapters we have discussed the basic transducers available for length, temperature and optical radiation, the three most important fields. In doing so we have automatically covered very nearly all the transducers there are. For example, although we have not specifically discussed force measurement, this is usually done by applying the force to some elastic member and measuring the resulting deflection with a displacement transducer, such as a strain gauge or an LVDT, and we have discussed these latter devices in some detail. It is arguable whether the measurement of quantities such as, say, force or acceleration should really be included in a book on transducers. Although a transduction action clearly occurs, such devices involve a combination of transducers — a mechanical modifier and a mechanical–electrical modulator in the case of force measurement — and the process is really more one of measurement than transduction. However, for completeness and to illustrate some of the applications of the basic transducers, we will briefly consider 'transducers' for some other fields, namely acceleration, force, pressure and flow.

8.1 Acceleration transducers

We discussed transducers for relative displacement and velocity under the heading of length transducers, but we totally omitted acceleration transducers. There is a good reason for this — there aren't any! A *relative* transducer makes a measurement with respect to some reference; for example, a relative displacement transducer, such as an LVDT, has one end fixed to an unmoving base. It happens that there are no simple transducers that respond directly to relative acceleration, but surprisingly absolute acceleration can be measured quite conveniently. An *absolute* transducer is one that does not require a reference; for example, if one wants to know the accelerations on a driver in a motocross car, one has to use a device that does not require a reference, since clearly no reference (a fixed surface) is available.

Acceleration transducers use what is known as the principle of seismic mass. They comprise a mass suspended by a spring from a rigid frame; the relative motion of the mass with respect to the rigid frame can be measured with a displacement or velocity transducer and is related to the acceleration of the frame. Figure 8.1 shows a simple accelerometer and its electrical analogue, from which one can easily deduce the transfer function.

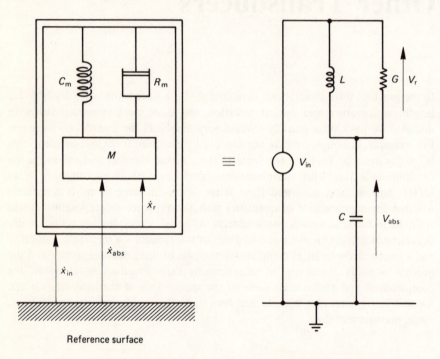

Figure 8.1 *Schematic accelerometer and its electrical analogue*

The absolute velocity of the rigid frame is $\dot{x}_{in}$ and the mass has relative velocity $\dot{x}_r$ and absolute velocity (which we do not require here) of $\dot{x}_{abs}$. In the analogue circuit the absolute velocity of the mass is represented by the voltage V_{abs} across the capacitor and the relative velocity is analogous to the voltage V_r across the parallel combination of inductance and conductance.

In terms of the Laplace variable s

$$\frac{V_r}{V_{in}}(s) = \frac{\dfrac{1}{\dfrac{1}{SL} + G}}{\dfrac{1}{\dfrac{1}{sL} + G} + \dfrac{1}{sC}} = \frac{s^2 LC}{s^2 LC + 1 + sLG}$$

Converting back to mechanical quantities and setting $\omega_0^2 = 1/mC_m$ and $\zeta = 1/2mR_m\omega_0$

$$\frac{\dot{x}_r}{\dot{x}_{in}}(s) = \frac{s^2}{s^2 + 2\zeta\omega_0 s + \omega_0^2} \qquad (8.1)$$

ω_0 is the natural angular resonant frequency and ζ the damping ratio. Equation (8.1) can be rewritten in terms of relative displacement x_r and input acceleration $\ddot{x}_{in}$ and becomes

$$\frac{x_r}{\ddot{x}_{in}}(s) = \frac{1}{s^2 + 2\zeta\omega_0 s + \omega_0^2} \qquad (8.2)$$

Equations (8.1) and (8.2) tell us all the important principles of the design of instruments for measuring absolute acceleration, and absolute velocity and displacement as well. The frequency response of the system of figure 8.1 is obtained by putting $s = j\omega$, and is shown in figure 8.2.

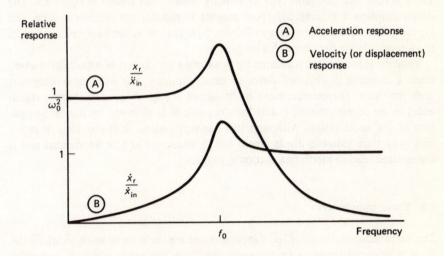

Figure 8.2 *Responses of accelerometer*

If we measure the relative displacement of the mass we get an accelerometer that has a flat response (of value $1/\omega_0^2$ up to the resonant frequency (curve Ⓐ). If we measure the relative velocity of the mass we get a velocity meter which has a flat response (of unity) above the resonant frequency (curve Ⓑ). Alternatively since $\dot{x}_r/\dot{x}_{in} = x_r/x_{in}$, curve Ⓑ shows we can get a measurement of absolute displacement above the resonant frequency by measuring the relative displacement of the mass. The main types of instruments are summarised below.

8.1.1   Accelerometers

All devices are of the same form in principle, the relative displacement of a mass/spring system being measured by a suitable transducer. The actual design depends very much on the frequency range required, since the response is determined by the natural frequency. At low frequencies (0–100 Hz) an LVDT or capacitive transducer is used as in figure 8.3(a). For higher frequencies (0–10 kHz) a cantilever beam arrangement with strain gauges is convenient (figure 8.3(b)) and for very high frequencies (up to 100 kHz) the mass is supported on the same piezocrystal which senses its motion, as shown in figure 8.3(c). Piezo-accelerometers can be very small but have the disadvantage that they do not respond below a few tens of Hz since piezo-electric displacement transducers cannot respond at d.c.

8.1.2   Velocity meters and displacement meters

There is only one common type of velocity meter, that shown in figure 8.4. The device employs a moving coil/fixed magnet transducer for velocity sensing, and typically has a natural frequency of about 5 Hz (as low as can easily be obtained), the response being flat above this frequency.

Velocity is most often measured by integrating the output of an accelerometer, when a response is obtained down to zero frequency, but with some integrator drift problems. Displacement can be measured at high frequencies (above resonance) by an accelerometer, but at low frequencies is obtained by double integration of the acceleration. Although this has very serious drift problems, it is the only way that absolute displacement can be measured at low frequencies and is the method used in inertial navigation systems.

8.2   Force transducers

The three main methods of force measurement are the mass balance, in which the force is balanced against a known mass, the force balance in which the balancing force is via a spring or magnet/coil arrangement and the deflection type, in which

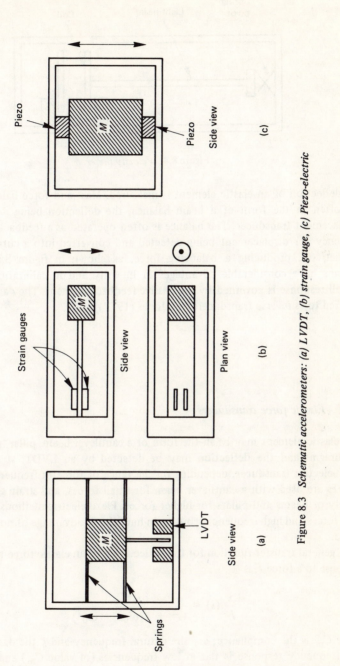

Figure 8.3 *Schematic accelerometers: (a) LVDT, (b) strain gauge, (c) Piezo-electric*

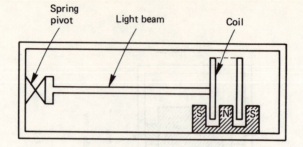

Figure 8.4 *Velocity meter*

the deflection of an elastic element is measured. Mass and force balance systems are often in the form of a beam balance, the deflection being detected by a displacement transducer. The balance is often operated as a feedback system, the tendency for displacement being detected and converted into a current fed to a magnet/coil, producing a balancing force, as shown in figure 8.5. Feedback systems have considerable advantages in linearity and in calibration, since the overall response is governed by the passive feedback element. They are sometimes referred to as *inverse* transducers (see Jones (1977)).

8.2.1 *Elastic force transducers*

The elastic element may be in the form of a cantilever beam, pillar, proof ring or diaphragm, and the deflection may be detected by an LVDT, strain gauge or piezo-electric transducer, depending on the range of force or frequency required. LVDTs are used with a cantilever beam for small forces, and strain gauges with a variety of beams and pillars for higher forces. Piezo-electric methods are used for high forces and high frequencies, but again have the disadvantage of not responding at d.c.

A general transfer function for the deflection x of an elastic force transducer in response to a force F is

$$\frac{x}{F}(s) = \frac{\omega_0{}^2 C_m}{(s^2 + 2\zeta\omega_0 s + \omega_0{}^2)}$$

where C_m is the compliance, ω_0 the natural frequency and ζ the damping ratio. The frequency response is flat at low frequencies (of value C_m) and falls off as $1/f^2$ above resonance.

Some typical force transducers are shown in figure 8.6.

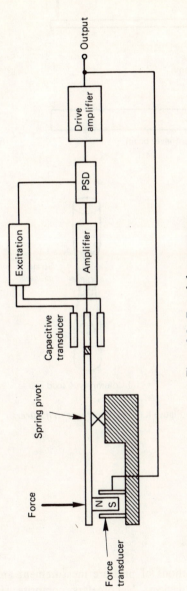

Figure 8.5 *Force balance system*

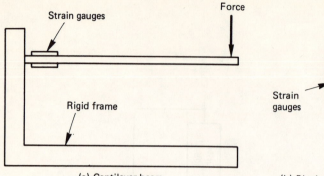

(a) Cantilever beam

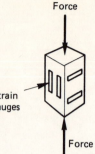

(b) Block-type load cell

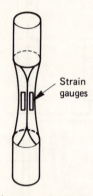

(c) Column-type load cell

Figure 8.6 *Various force transducers*

8.3 Pressure transducers

The most common methods of pressure measurement are by manometers and mechanical deflection devices. Manometers comprise U-tubes with one end closed, usually containing mercury, the difference in height in the two sides being proportional to the pressure applied at the open end. Deflection transducers may be tubes, diaphragms or bellows, the deflection being measured by a suitable displacement transducer.

8.3.1 Elastic pressure transducers

Bourdon tubes have been widely used for pressure measurement. They consist of a flattened tube of approximately elliptical cross-section bent into some shape, such as a 'C' or a spiral, so that the end deflects under pressure. Some examples are shown in figure 8.7. They are fairly linear for small deflections. Bellows are also used and have an improved range and linearity; they are reversible and are useful as displacement-pressure transducers in pneumatic systems.

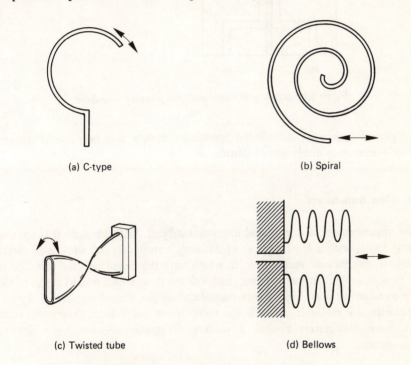

(a) C-type   (b) Spiral

(c) Twisted tube   (d) Bellows

Figure 8.7 *Various elastic pressure transducers*

Diaphragms are probably the most popular transducer. They consist of a thin stretched membrane whose deflection is usually measured by capacitive or inductive methods, though strain gauges may be used. The deflection of the centre of the membrane, x, is linearly dependent on the pressure applied, p, for deflections up to about half of its thickness, and is given as

$$x = \frac{3(1 - v^2)d^4 p}{256Et^3}$$

where v = Poisson's ratio, d = diameter, E is Young's modulus and t is the thickness.

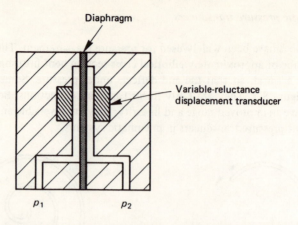

Figure 8.8   *Differential diaphragm-type pressure transducer*

Figure 8.8 shows a differential transducer with a soft iron diaphragm and variable-reluctance displacement transducer.

8.4   Flow transducers

Flow measurement is a large and important subject. Unfortunately it is a rather messy topic, with a large number of different transducers but with little overall unity. It may include vector flow, in which both the magnitude and direction of the flow at a point in space are required, volume flow, which usually refers to the flow in tubes, or mass flow, where the total mass per second is required. We will summarise the methods available for vector flow and volume flow only, since most mass flowmeters involve a volume flowmeter followed by a density measurement.

8.4.1   *Vector flow transducers*

Apart from various forms of probes and vanes for measuring the direction of flow, the best-known methods are by means of Pitot tubes and anemometers. A Pitot tube is shown in figure 8.9. It involves a hollow tube feeding an inclined manometer, and has to be oriented to give a maximal reading. Pitot tubes are often used for volume flow measurement in pipes, especially when the profile across a section is needed.

The hot-wire anemometer comprises an electrically heated wire placed in the fluid. The amount of cooling depends on the direction and rate of flow and may be detected by temperature transducers. Alternatively the wire may have a known resistance/temperature characteristic and can form one arm of a bridge circuit.

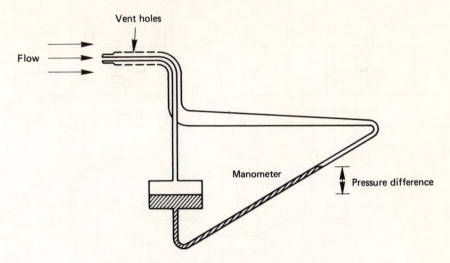

Figure 8.9 *Pitot tube flow transducer*

The heating current may be maintained constant, but an alternative method is to employ a feedback system to maintain the wire at constant temperature, so the heating current becomes proportional to the flow.

8.4.2 *Volume flow transducers*

In measuring the flow of a liquid through a pipe, an ideal transducer would be one that causes no disturbance to the flow (known as a 'non-invasive transducer') and that preferably can be situated externally to the pipe. Unfortunately very few devices fulfill either of these requirements, though some of the more recent methods do so under certain conditions.

Most flow transducers are simply modifers and comprise plates or nozzles which partly restrict the flow and produce a pressure drop. This is exactly analogous to the measurement of electric current by finding the voltage drop across a small series resistor. Some typical devices are shown in figure 8.10. In each case the volume per second is proportional to the square root of the pressure difference, which is measured by a suitable transducer such as that shown in figure 8.8 above.

These transducers are far from ideal, requiring a special insert in the pipe, but are nevertheless very widely used.

Rotameters and turbines are rather more satisfactory. A rotameter is a shaped float whose height in a vertical tube containing the fluid depends on the flow rate. A turbine is a suitably vaned device placed directly in the flow. It has the advantages that the disturbance is small, and it may work bidirectionally. In addition, the rate of rotation can be sensed externally to the tube and may be

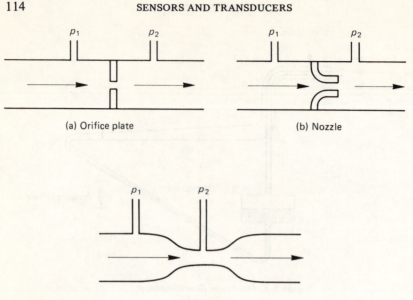

(a) Orifice plate

(b) Nozzle

(c) Venturi

Figure 8.10   *Various pressure-drop flow transducers*

read directly in digital form. The drag-force flowmeter is somewhat similar in principle, being a beam placed perpendicular to the flow, which is deflected and its displacement measured by strain gauges or an LVDT.

The electromagnetic flowmeter is very satisfactory for conducting fluids, and comprises two electrodes in a non-conducting section of tube, as shown in figure 8.11. A strong magnetic field is applied perpendicularly to the plane of the diagram, and the ions are deflected according to their sign and the fluid velocity, and detected by the electrodes placed in a suitable bridge circuit. Apart from the inserted section of tube, this is almost an ideal transducer.

Ultrasonic measurement techniques are increasing in popularity and are easily applied to flow measurement, though their accuracy depends on the fluid concerned and is low for fluids of low density. The principle is that disturbances are propagated from a transmitter through the fluid to a receiver. The transit time will be dependent on the component of fluid flow in the direction of propagation. For suitable fluids these methods meet both the requirements of the ideal transducer. A good review is given by Asher (1983).

Doppler methods also have this advantage, though they are applicable only to transparent fluids. Figure 8.12 is a schematic diagram of a laser doppler flowmeter. A powerful main beam and a weaker reference beam are passed through the fluid at different angles, and some of the main beam is reflected in the direction of the reference beam by any small eddies or particles in the fluid. These act as doppler reflectors so there is a slight frequency shift, proportional to the component of velocity in the direction of the main beam. Because optical frequencies are

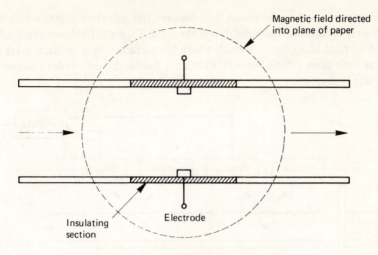

Figure 8.11 *Electromagentic flowmeter*

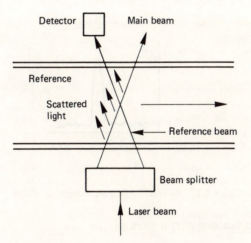

Figure 8.12 *Laser doppler flowmeter*

so high, a beat frequency can be counted electronically and flows over a wide range (a few mm/s to several hundred m/s) measured.

An interesting recent method, which has ideal characteristics and wide application, is flow measurement by correlation. The principle is that a beam of light is reflected from two adjacent points in the flow (alternatively it may be transmitted) and the correlation function for the two signals $s_1(t)$ and $s_2(t)$ are determined. Provided that the two reflection points are reasonably close, $s_2(t)$ will be simply a delayed version of $s_1(t)$ and the correlation function will have a peak for a time interval equal to the transit time between the reflection points, which is of course proportional to the velocity of the fluid. Figure 8.13 shows the method applied to fluid in an open channel, where the surface ripples produce $s_1(t)$ and $s_2(t)$; in some cases a disturbance (for example, bubbles) must be added to provide something to correlate.

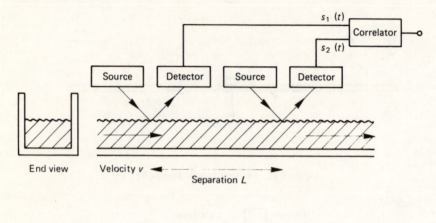

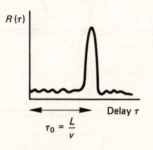

Figure 8.13 *Correlation flow-measurement system*

The correlation function $R(\tau)$ is given by

$$R(\tau) = \int_{-\infty}^{+\infty} s_1(t)s_2(t + \tau) \; dt$$

where τ is a time shift. This is not very convenient for calculation but a considerable simplification is obtained by digitising $s_1(t)$ and $s_2(t)$ into binary values. The correlation, though with some reduction in accuracy, can be found rapidly by an on-line microprocessor, since it involves only EOR (exclusive-OR) operations. Indeed, the method became feasible only with the developments of cheap processors, but can now be applied in many fields such as for measuring the speeds of belts in a machine, the speed of cars or trains and even the velocity of effluents from remote factory chimneys. Beck (1983) has given a useful review.

8.5 Exercises on chapter 8

8.5.1. Use the general transfer junction of an accelerometer (equations (8.1) and (8.2)) to deduce the types of instrument required for the following measurements.

(i) Absolute velocity over the range 10 Hz to 1 kHz

(ii) Absolute acceleration over the range 10 Hz to 10 kHz

(iii) Absolute displacement over the range 0 to 100 Hz.

8.5.2. Dicusss the methods available for the measurement of rotational and translational velocity of an object, distinguishing between absolute and relative methods. Illustrate your answer by reference to the measurements below.

(i) Measurement of the speed of rotation (near 100 rev/s) of a shaft by a non-contacting method with an accuracy of 0.1 per cent.

(ii) Measurement of the relative velocity between two parts of a machine moving in translation, with a range of 2 cm and a maximum velocity of 10 m/s.

(iii) Measurement of the absolute vertical velocity of a concrete block on the floor of a workshop. The vibrations of interest are in the range 5–100 Hz and have maximum amplitude 0.1 mm.

(iv) Measurement of the speed of a belt in a machine by a remote method with an accuracy of 10 per cent. The belt has a nominal speed of 10 m/s and is opaque and non-conducting. A readout is required every 10 s.

8.5.3. Figure 8.14(a) shows a strain gauge accelerometer, which consists of a mass of 0.1 kg attached to a beam of mild steel of elastic modulus $E = 2 \times 10^{11}$ N/m^2. A strain gauge is attached to each side of the beam and connected differentially in the bridge circuit of figure 8.14(b). Each gauge has an unstrained resistance of 1 kΩ and a gauge factor of 2.

(i) Find the natural undamped frequency of the device (assume that the mass of the beam is negligible).

(ii) Find the zero-frequency acceleration responsivity at the amplifier output (assume that the beam is critically damped).

(iii) Find the maximum input acceleration, assuming this is set by the amplifier limiting when its output reaches ±10 V.

Note: A force f applied to the end of the beam produces a deflection at the free end of $4fl^3/(Ebd^3)$ and a surface stress at distance x from the free end of $6fx/bd^2$.

8.5.4. Figure 8.15 shows a schematic piezo-electric accelerometer in which a mass of 0.05 kg is constrained to move horizontally by two quartz crystals,

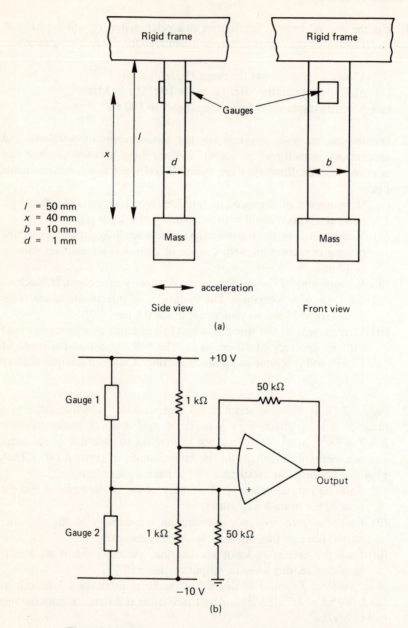

Figure 8.14   *Strain gauge accelerometer (a) and amplifier (b)*

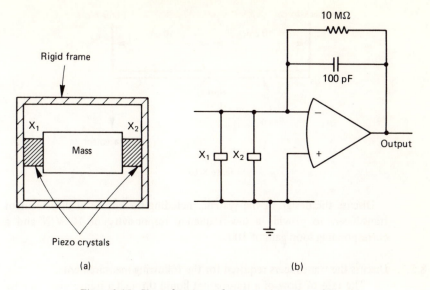

Figure 8.15 *Piezo-electric accelerometer (a) and amplifier (b)*

each of area 1 cm² and thickness 0.5 mm. The signals from the crystals feed an operational amplifier.

Find the natural undamped frequency of the mass and the mid-frequency responsivity at the amplifier output, and sketch the responsivity versus frequency.

[Quartz has an elastic modulus of 10^{11} N/m², a dielectric constant of 4.5 and a charge/force coefficient of 2×10^{-12} C/N.]

8.5.5. A d.c. force measurement system is required for forces up to 1000 N using a cantilever beam clamped at one end with the force applied to the free end. The cantilever is of mild steel and of length 5 cm, width 2 cm and thickness 4 mm. Discuss the advantages and disadvantages of the transducers below, and recommend the most suitable type.
(i)   Strain gauges
(ii) Piezo-electric crystals
(iii) Variable-separation capacitive transducers.
[Use the equations given in section 8.5.3.]

8.5.6. A force-feedback weighing machine comprises a cantilever beam supported by spring pivots as shown in figure 8.16. The beam is uniform in section and has a length of 20 cm, a mass of 0.1 kg and a natural frequency of 5 Hz. The force transducer has a force/current coefficient of 10 N/A.

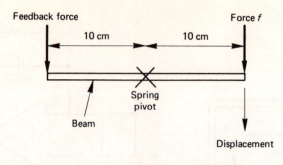

Figure 8.16

Discuss the design of the system, including the choice of displacement transducer, to provide a low-frequency responsivity of 10 V/N and a corresponding loop gain of 100.

8.5.7. Discuss the transducers required for the following measurements.

   (i) The rate of flow of a transparent liquid through a transparent pipe by a non-contacting method (bubbles or other disturbances may be assumed to be present).

   (ii) The rate of flow of water through an open channel.

   (iii) The flow velocity at various positions and depths in a river.

9

Recent Developments in Transducers

The technological advances of the last few years have led to an increased interest in transducers, particularly in devices suitable for direct interfacing to microprocessor systems. Most fields of measurement have progressed rapidly, but we will concentrate here on simple transducers rather than transducer systems (for force measurements, flow measurement etc.). Three areas stand out: optical-fibre transducers, resonator sensors and solid-state transducers. In each case important advances have already been made, but there is an exciting potential for developments in the next few years.

9.1 Optical-fibre transducers

The considerable developments in optical fibres over the last decade were primarily aimed at the communications field, of course, and it is only over the last few years that the possibilities for transducer developments have become apparent. An optical fibre is a thin low-loss fibre in which total internal reflection of a beam of light entering at one end causes the beam to be contained completely within the fibre. Fibres may be very thin monomode types requiring a laser source or thicker multimode types which may be used with an LED. There are two distinct ways in which optical fibres may be used for sensing physical (or chemical) quantities. They can be used simply as signal guides to and from the area of interest, such as an oven in a temperature-measurement system or a moving vane in a displacement system. Alternatively, the light passing through the fibre can be affected in some way by an external parameter; for example, the refractive index changes if the fibre is stressed. The particular advantages offered by optical fibres in transduction are their inherent immunity to electromagnetic interference and the ease with which they can be linked to an optical communication system.

121

The properties of light available for modulation in sensing applications are intensity, phase, polarisation, wavelength and spectral distribution. All these properties have already been used in sensors, though most devices use the first three. In terms of the forms of energy discussed in chapter 1 we can find modulating effects in most cases, as shown in table 9.1, though of course there are no self-generators. However, in looking at examples of various transducers, it is easier to group them by the property of light that they employ.

Table 9.1　Modulating effects in optical fibres

Energy type	Modulating effects	Physical effect
Mechanical	Stress birefringence	Refractive index and absorption
	Piezo-absorption	Absorption
Electrical	Electro-optic effect	Refractive index
Magnetic	Magneto-optic	Refractive index
	Faraday effect	Polarisation
Temperature	Thermal	Refractive index and absorption

Intensity modulators

Both signal-guiding and external-parameter devices exist. Figure 9.1(a), (b) and (c) shows examples of the former type, whereas figure 9.1(d) shows a 'microbending' system. By using clamping plates with a periodic mechanical grating it is possible to measure a number of parameters such as displacement, pressure, strain and temperature.

Phase modulators

Phase changes can be produced by mechanical strain, temperature etc., but the effect is mostly used for more rapidly changing parameters such as acoustic fields in a recently developed hydrophone. Interferometric methods are used, with a coherent source. One of the most common schemes is the Mach–Zender interferometer in figure 9.2(a). A monomode fibre is used with a modulator, usually a Bragg cell, which causes a suitable difference frequency to be produced, at which the phase shift is detected. A simpler method using a multimode fibre, in which phase modulation occurs between different modes in the same fibre, is shown in figure 9.2(b). An important application is the fibre-optic gyroscope; light is directed in opposite directions in a coil of fibre and a small phase shift is produced on rotation.

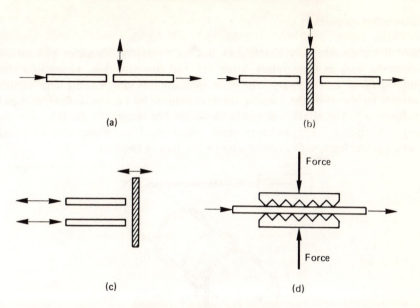

Figure 9.1 *Intensity modulators*

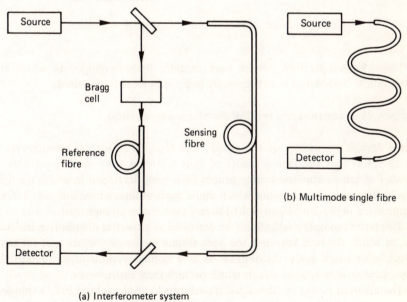

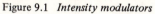

(a) Interferometer system
(monomode fibre)

(b) Multimode single fibre

Figure 9.2 *Phase modulators*

Polarisation modulators

Most fibres are internally birefringent, but unfortunately this varies with various conditions such as temperature, stress etc. and also with time, so specially constructed fibres are necessary. The main application is in measuring large electric currents by means of the Faraday rotation induced by a magnetic field, as shown in figure 9.3. The effect is proportional to the line integral of the field over the length of the fibre and, although small, the method has proved useful for high voltage power lines, with a resolution of a few tens of ampères.

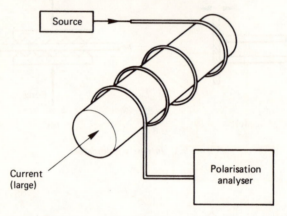

Figure 9.3   *Polarisation modulator for current measurement*

Miniature temperature probes have recently been developed in which the temperature dependence of orthogonally polarised modes is exploited.

Wavelength modulators and spectral distribution modulators

These devices are mostly of the signal-guide type. The Doppler anemometer in figure 9.4(a) uses the Doppler shift of light scattered from moving particles. A number of temperature-measuring probes have been developed in which the light is guided to a special phosphor which emits spectral lines whose intensity ratio is temperature dependent; figure 9.4(b) shows a schematic arrangement.

This latter example could almost be described as a spectral distribution modulator, of which the best known is the optical-fibre pyrometer, which uses the same principle of black body distribution as in a normal pyrometer but avoids the unpredictable atmospheric effects which perturb these instruments.

The inherent power of fibre-optic transducers has not yet been fully exploited. They are not ideally suited to interfacing to most microprocessors, though opto-electric conversion is fast and efficient. They do not at present utilise the enormous information capacity of optical fibres, nor do they make use of the ability

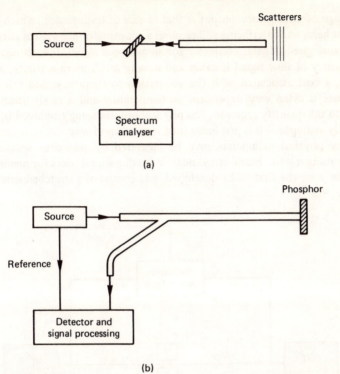

Figure 9.4   *Wavelength and spectral distribution modulators*

of optical systems to perform 'instantaneous' Fourier transforms. However, it is clear that these features will be developed in the future, when multiplexed transducers, parallel data transmission and 'instantaneous' processing will offer remarkable possibilities. If 'optical computers' are developed they would of course be ideally suitable. Recent reviews of optical-fibre transducers have been given by Culshaw (1982) and Harmer (1982).

9.2 Resonator sensors

Resonator sensors are mechanical modifiers in which a mechanical element is excited into vibration at its natural frequency, the value of which depends on the desired input quantity. The output is thus a frequency proportional to the quantity of interest.

The advantages usually claimed of a frequency output, as opposed to an analogue voltage, are ease of interfacing to digital processors, stability, freedom from electrical interference and low power requirements. Actually the main advantage of a frequency output is that of ease of transmission, which is not very relevant here, and interfacing still requires a counting/timing system corresponding to an analogue-to-digital converter for an analogue signal, though digitisation of a frequency or time signal is easier and usually much more accurate. There is, of cource, a cost associated with the conversion to frequency, and this is that the resonance is often very dependent on temperature and is rarely linearly proportional to the quantity required. The physical process being employed is, of course, perfectly analogue — it is just being used in a different way.

Many physical parameters may be measured by resonator sensor methods, usually using a wire, beam or cylinder. Vibrating-wire devices for measuring force or strain were the first to be developed, and consist of a stretched wire, as shown in figure 9.5.

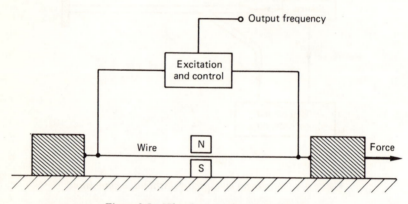

Figure 9.5   *Vibrating-wire force transducer*

The resonant frequency f_r of a wire of length l and mass per unit length m, stretched by a force T is

$$f_r = \frac{1}{2l} \sqrt{\frac{T}{m}}$$

Wires of tungsten or indium are usually best. Excitation is by means of a powerful permanent magnet and a driver circuit which feeds current through the wire, and maintains the frequency at resonance by sensing changes in the electrical impedance of the vibrating wire. Some compact and accurate pressure transducers have been developed recently in which the wire is attached to a diaphragm, though considerable processing is required to remove temperature effects and produce a linear output.

Vibrating beams follow similar principles to vibrating wires, being used mostly for force, though other parameters can be measured. Figure 9.6 shows a beam used in flexure for level measurement, a similar arrangement for flow, a beam vibrating along its length for viscosity measurement and a beam in torsional vibration for density measurement.

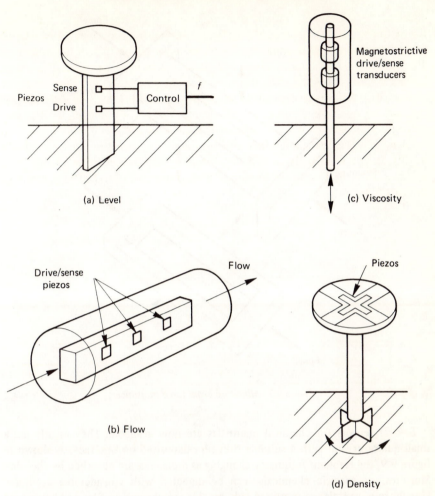

Figure 9.6 *Vibrating-beam transducers*

Resonating beams for force measurement are often in the form of two tuning forks connected tine-to-tine, as shown in figure 9.7, providing very high Q-factors. They may be made very small, and some of the most recent devices are of quartz or even etched from silicon, in which case the dimensions may be less than 1 mm. The mechanical design is very important since it must be possible to excite a single vibratory mode of high Q-factor, and thermal effects must be minimised.

Vibrating cylinders and tubes are also popular. Excitation of a suitable mode is more difficult and substantial processing is required to obtain a useful output, but several different quantities may be measured in this way. Figure 9.8 shows a 'tuning fork' arrangement for density measurement and a tube in torsional vibration for detecting mass flow.

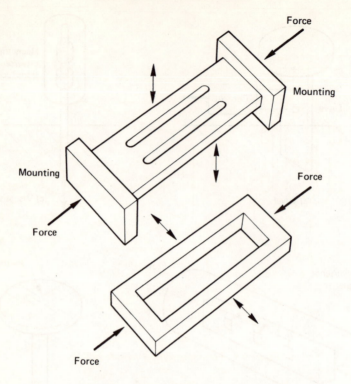

Figure 9.7 *Miniature beam force transducers*

Several sensors for chemical quantities are now available. They mostly use a small quartz crystal with a suitable thin film deposited on one face, as shown in figure 9.9, the resonant frequency changing as molecules are absorbed by the film. Moisture, or specific chemicals, can be detected with considerable accuracy. Usually two crystals are mounted side by side, one having no film and being used for reference purposes. Mass changes at the microgramme level can be detected.

9.3 Solid-state transducers

Most solid-state transducers are based on silicon. It has been said that silicon is not an ideal material for sensors, but nevertheless it has a number of excellent properties. The most important are the ease with which it can be purified to almost unbelievable levels and the fact that its oxide, produced by heating in an oxygen atmosphere, forms an excellent mask for the selective doping and etching of a silicon-based structure. Most of our present technology depends on these properties, of course, but similar possibilities clearly exist for the development of transducing devices employing essentially the same fabrication techniques.

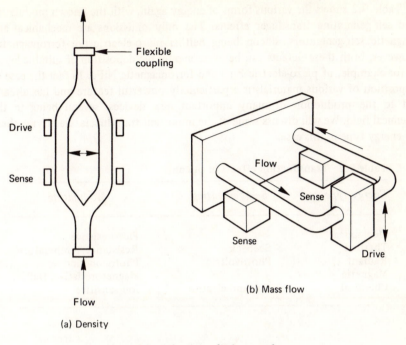

(a) Density

(b) Mass flow

Figure 9.8 *Vibrating-cylinder transducers*

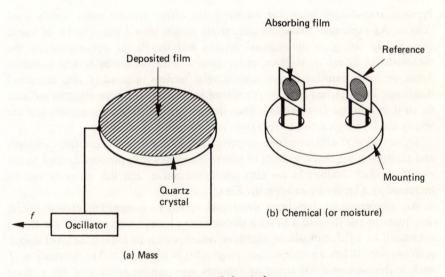

(a) Mass

(b) Chemical (or moisture)

Figure 9.9 *Mass and chemical sensors*

Table 9.2 shows the various forms of energy again, with the known modulating and self-generating transducer effects. The only omissions are mechanical and magnetic self-generators, silicon being neither piezo-electric nor ferromagnetic. However, both these defects can be overcome by the deposition of suitable layers — for example, of piezo-electric ZnO and ferromagnetic NiFe. In fact the ease of deposition of various materials is a particularly powerful feature and has already led to the production of many important new devices, several being in the chemical field. We will discuss some of the important transducers, classifying them by energy type in this case.

Table 9.2   Transduction effects in silicon

Energy type	Self-generating	Modulating
Mechanical	—	Piezoresistive
Thermal	Seebeck	Resistance/temperature
Radiant	Photovoltaic	Photoconductive
Magnetic	—	Magnetoresistive, Hall
Chemical	Galvano-electric	Ion-sensitive

Mechanical transducers

Pressure transducers using the piezoresistive effect are the most widely used devices. As explained above, metallic strain gauges have a gauge factor of about two, purely owing to dimensional effects, but this is far outweighed by the piezoresistive effect in semiconductor gauges, with factors of several hundred. Some pressure transducers are now integral devices, a small (1 mm diameter) diaphragm being etched from n-type silicon and with four p-type resistors diffused on to it in a bridge arrangement. Such devices are available commercially and are widely used, though active temperature compensation is necessary.

Other physical effects usable in pressure transducers are capacitance changes and changes in the characteristics of planar transistors with pressure applied to the emitter surface. MOSFETs are also pressure-sensitive, and this property can be increased by a layer of piezo-electric ZnO.

An accelerometer has been developed recently, comprising etched silicon structures in the form of a minute silicon mass (1 mm^2 area and 200 μm thick) connected to a 0.5 mm silicon cantilever beam, which includes a diffused silicon piezoresistor. It has an acceleration range of 0.1–1000 ms^{-2}. The fabrication of such a three-dimensional structure requires very precise control of the etching processes, but the excellent properties of silicon have permitted the development of various devices of this form.

The piezoresistive effect has also been used in anemometers, in which differential cooling of diffused silicon resistors in a bridge arrangement gives a measure of the rate of flow of air.

Thermal transducers

The bandgap in silicon is comparatively large, so doping is necessary to produce useful temperature effects (that is, silicon behaves as an extrinsic semiconductor at room temperature). A few resistance thermometers have been developed, mostly for rather specialised applications. However, the most widely used effect is the temperature-dependent voltage across a forward-biased diode, discussed above. The linearity and stability are not high, but the possibility exists of adding some signal-processing elements on the same chip and much development is taking place in this direction.

Another approach is to use two transistors, usually a dual transistor, operated at a fixed emitter/current ratio, when it is found that the difference in base-emitter voltages is proportional to absolute temperature. Again it is necessary to include some processing circuitry, and a compact accurate device can be fabricated.

Radiation transducers

Photoconductors and photodiodes are the best-known examples of silicon radiation transducers, and have been described in some detail in chapter 7. Charge-coupled devices (CCDs) are also widely used and are attracting much attention because of the possibility of producing solid-state television cameras. They utilise the dependence of capacitance on voltage in MOS diodes, and consist of an array of diodes in the form of a shift register. Light falling on the device generates packets of charge, which can be manipulated by suitable applied voltages. Parallel arrangements of large numbers of such registers enable two-dimensional images to be analysed, and arrays of up to a million elements have been produced.

Magnetic transducers

Since silicon is not magnetic, only modulating magnetic transducers are available; these mostly employ either the Hall effect or the magnetoresistive effect discussed in chapter 3. Silicon has a high Hall coefficient, and thin Hall plates are available, comprising resistive *n*-type layers on a *p*-type substrate. They may be used for the measurement of fairly large magnetic fields or electric current (with a fixed magnetic field), and often include some signal processing on the same chip.

The magnetoresistive effect in silicon is fairly small, so silicon is usually used as a substrate for more suitable materials, such as indium antimonide. Films of ferromagnetic NiFe are also used, and can exhibit significant changes in resistance owing to rotation of the direction of magnetisation.

The characteristics of diodes and transistors show a dependence on magnetic force. The most suitable arrangements comprise some form of dual transistor, the difference in collector currents being proportional to field strength. The devices are also sensitive to field direction.

Chemical transducers

In the past, most sensors for chemical quantities have tended to be rather bulky and complicated, but the area has received a large impetus with the development of various miniature solid-state sensors. The first device was the ISFET (ion-sensitive FET), comprising a MOSFET with an ion-selective layer of a polymer (Si_3N_4, Al_2O_3) grown on the gate. The characteristics are affected when immersed in an electrolyte, and the pH can be measured. The palladium-gate MOSFET was later developed, the replacement of the normal gate by palladium leading to a sensitivity to hydrogen gas.

A range of transducers has been produced in which a suitable polymer is deposited on an electrode structure or on the gate of silicon MOSFETs. Devices sensitive to CO, CO_2, CH_4, SO_2 and NH_3 have been made. They are usually sensitive to moisture content as well, and various humidity sensors are available. Another form of humidity sensor involves an electrode structure for detecting capacitance changes caused by dew formation, together with a semiconductor temperature sensor and miniature Peltier cooling element. It is really a miniature version of a conventional humidity sensor, using silicon as the substrate. A similar technique is used in an oxygen sensor, using etched silicon channels containing an electrolyte of NaCl, the cell current being proportional to the pressure of oxygen diffusing through a suitable membrane.

Other materials

Various other materials have been used for solid-state sensors, notably germanium and gallium arsenide. The latter is considered to have great potential, but does not have the advantage of silicon, whose oxide forms such an excellent mask, making large-scale integration less convenient.

A particularly promising material for piezo-electric and pyro-electric sensors is polyvinylidene fluoride (PVDF). Most piezo-electric materials have very high stiffness, limiting their applications to pressure and force transducers, but PVDF is available in thin films of low stiffness and also relatively low permittivity, and will find considerable application in the next few years.

A recent review of silicon transducers has been given by Middlehoek and Noorlag (1981a).

9.4 Smart sensors

Although the word 'smart' is somewhat misused here, the meaning is reasonably clear. The term refers to transducers in which intelligence is added by means of a digital processor, preferably on the same chip. Silicon sensors are ideally suited, of course, and the piezoresistive, photovoltaic and Hall effects are particularly useful.

A general scheme has been suggested by Brignell and Dorey (1983), and is shown in figure 9.10.

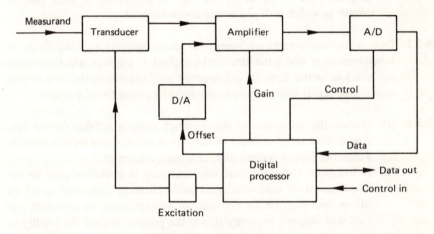

Figure 9.10 *Smart sensor*

The transducer is under total processor control. The excitation is produced and modified if necessary, depending on the range required. The offset (of the transducer or amplifier) is controlled, as is the gain, and the A/D is similarly adjusted. The processor contains full details of the transducer characteristics in ROM, enabling the correct excitation and gain etc. to be maintained under all conditions. It also contains details of responsivity, including non-linear behaviour, so that the final output is always an accurate transduction of the measurand. Indeed, it is already becoming accepted that non-linearity and offsets in transducers are no longer serious shortcomings.

In particular, it is apparent that the actual type of transducer in figure 9.10 is of little importance. Its physical characteristics need to be known and understood, and the basic transducer responsivity and detectivity still determine the overall detectivity. However, whether or not the transducer is actually a so-called digital type is largely irrelevant. The processing would be slightly different and the A/D would be replaced by a scaling device, but all the processing and control would be invisible to the user. As we pointed out before, nearly all transducers are really analogue anyway, but sometimes it is convenient to operate them in a digital manner. We are analogue animals in an analogue world, which is where we started in chapter 1, except that the writer's two word-processing fingers are a bit shorter now.

9.5　Exercises on chapter 9

9.5.1. (i) Discuss the properties of optical fibres which may be used for sensing purposes, and the advantages and disadvantages of such systems.

(ii) Explain what is meant by the optical processing of data and the extent to which such processing is now possible.

9.5.2. Discuss the advantages of resonator sensor systems and the fields of measurement in which the idea can be applied. Is a system which produces information in the form of a frequency proportional to the desired parameter necessarily superior to one producing a proportional voltage?

9.5.3. (i) Discuss the properties of silicon which make is suitable for the construction of large numbers of very small circuits, and its suitability as a material for the construction of sensing elements.

(ii) It has been suggested that the next step in evolution may be the development of super-intelligent self-replicating 'creatures' based on silicon, and that the development of organic creatures culminating in man was simply a necessary step in the process. Discuss the validity of the premises on which the theory is based.

10

Solutions to Exercises

10.1 Solutions to exercises on chapter 1

1.6.1. *Responsivity* is response per unit input.
Detectivity is output signal-to-noise ratio per unit input.
Range is the range of input levels over which the responsivity satisfies some stated requirement.
 (a) An LVDT (linear variable differential transformer).
  Responsivity typically 100 V/m per volt excitation.
  Detectivity better than 10^6/m (< 1 μm change detectable).
  Range ±1 cm (non-linearity < 1 per cent over this range).
 (b) A resistance thermometer (Platinum type).
  Responsivity typically 0.5 Ω per 100 Ω per °C.
  Detectivity typically 1000/°C (limited by noise).
  Range -200 to $+ 850$°C (1 per cent non-linearity).
 (c) A photovoltaic cell (silicon solar cell).
  Responsivity typically 0.1 A/W (at 1 μm).
  Detectivity 10^{12}/W.
  Range 0 to 1 W (upper limit set by heating effects and no precise value can be given).

1.6.2. (a) (i) Modulator
   (ii) Self-generator
   (iii) Modulator
   (iv) Modifier (displacement–angle)
   (v) Self-generator (thermal–mechanical).
  (b) (i) Thermistor or resistance thermometer
   (ii) Piezo-electric crystal or tachometer
   (iii) Diaphragm-type barometer

135

(iv)   Silicon solar cell
(v)   LVDT or capacitive displacement transducer.

10.2   Solutions to exercises on chapter 2

2.6.1.

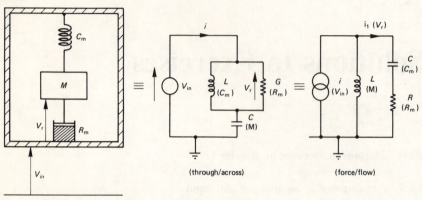

Figure 10.1 *Accelerometer and electrical analogues*

2.6.2. (i)

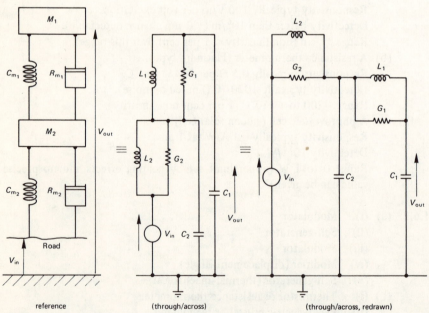

Figure 10.2 *Suspension system and electrical analogue*

The redrawn electrical circuit clearly shows that there is a two-stage filtering effect, so we require the inductors (compliances) to be large, the conductances (damping) small and the capacitors (masses) large.

(ii) C_1 (mass of car): say 500 kg. $C_1 = 500$ F.

C_2 (mass of wheel): say 25 kg. $C_2 = 25$ F.

L_1 (main suspension): Suppose the spring deflected by 30 cm in supporting the mass of the car

$$C = \frac{0.3}{500\,g} \times 6 \times 10^{-5}$$

so $L_1 = 6 \times 10^{-5}$ H.

L_2 (tyre compliance): Suppose tyre deflects by $2\frac{1}{2}$ cm in supporting the mass of the car

$$C = \frac{2.5 \times 10^{-2}}{500\,g} \approx 5 \times 10^{-6}$$

so $L = 5\ \mu$H.

Note that the C values are far too large for a direct model to be made, so in practice some scaling is necessary. The Q-factor is given by $Q = 2\pi M/R_m T$ where T is the period, so R_m can be deduced by measuring the rate of decay of oscillation.

2.6.3. This is a current-driven system (by the heater), feeding a parallel combination of resistance and capacitance. The ambient temperature can be assumed constant and equivalent to electrical ground. An equivalent circuit is shown in figure 10.3.

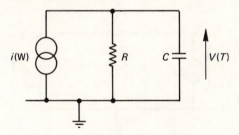

Figure 10.3

The transfer function is $\dfrac{T}{W}(f) = \dfrac{R}{1 + j\omega CR}$

R: 25°C temperature difference maintained by 2.5 W so $R = 10°$C/W.
C: $400 \times 0.025 = 10$ J/K.
Time constant $= RC = 100$ s.
The circuit is essentially a first order low-pass filter.

2.6.4.

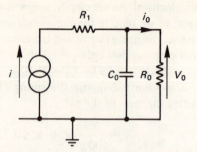

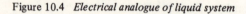

Figure 10.4 *Electrical analogue of liquid system*

R_1 and R_0 produce pressure drops (voltage drops).

2.6.5. (a) A lever (or a gearbox).

(b) Yes (all masses represented by capacitors with one end grounded).

(c) No (because some capacitors may have neither end grounded). It is possible to overcome this limitation, but only with considerable complication.

(d) None of them do (because there is only one storage element).

(e) It is known as 'inertance' and is proportional to liquid density, but is usually small and may often be ignored.

(f) The variables are pressure p and mass flow rate i (unlike liquid systems the volume is constant, so pressure must be used instead of height, and mass flow instead of volume flow). The elements are capacitance (volume) and resistance (restrictance). The inertance is always negligible for gases.

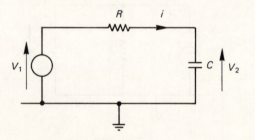

Figure 10.5 *Electrical analogue of gas system*

10.3 Solutions to exercises on chapter 3

3.5.1. (a) Types of energy (relevant to measurement): mechanical, electrical, magnetic, thermal, radiant.

Modifiers: convert between forms of same type of energy, usually between the through variable and the across variable, for example, flow transducer (flow → pressure)

Self-generators: transduce directly between two types of energy, for example, piezo-electric crystal (deformation → charge)

Modulators: require external energy source and modulate flow of energy, for example, thermistor in bridge circuit.

(b) Self-generators: radiant–electrical: photovoltaic effect; silicon solar cells.

mechanical–electrical
$$\begin{cases} \text{electromagnetic effect; tacho-generators} \\ \text{piezo-electric effect; force transducers etc.} \end{cases}$$

magnetic–electrical: electromagnetic effect; magnetic field measurement

thermal–electrical: thermo-electric effect; thermocouples.

Modulators: (electrical in, electrical out).

radiant: photoconductive effect; photoconductive cells

mechanical:
$$\begin{cases} \text{piezoresistive;} \\ \text{dimensional-resistive;} \end{cases}$$
strain gauges (semi-conductor and metallic)

magnetic: magnetic-resistive; magnetic fields (little used)

thermal: thermal-resistive; resistance thermometers, thermistors.

Modifiers: mechanical: flow/pressure; orifice-plate flow transducer
radiant: heat flux/temperature; thermal photodetector
thermal: heat flux/temperature; anemometer.

3.5.2. (i)

Photovoltaic.	Radiant flow–electric current.	Transforming
Pyro-electric.	Temperature–charge.	Gyrating
Electromagnetic.	Velocity–voltage.	Transforming
Piezo-electric.	Displacement–charge.	Gyrating
Thermo-electric.	Temperature–voltage.	Transforming

(ii) Mechanical

Beams and springs:	Force-displacement.	Gyrating
Diaphragms and tubes:	Pressure–displacement.	Gyrating
Orifices etc.:	Flow–pressure.	Gyrating
Radiant		
Photodetectors:	Radiation–temperature.	Gyrating

Thermal
Anemometers: Heat flow–temperature.   Gyrating
Note: There is no point in doing this for a modulator since it can be
arranged to produce either a current or voltage output as required.

3.5.3. Faraday's law ($e = - nd\phi/dt$) can be used directly in measuring changing
magnetic fields, as a magnetic–electric self-generator. However, it is more
useful for velocity measurement as a tachometer, involving a magnet/coil
system, and is then said to employ the electrodynamic effect. The effect
is then that of a mechanical–electrical self-generator (at first sight it appears
to be a magnetic–mechanical–electrical modulator, though in fact no
magnetic energy is used). A similar effect occurs in the electromagnetic
flowmeter, in which a magnetic field applied perpendicular to the flow of
a conductive fluid produces an e.m.f. across electrodes in the fluid.

Faraday's law also applies to the movement of individual charge carriers
in semiconductors and gives rise to the magneto-resistive effect, in which
deflection of the carriers in a magnetic field leads to an increased resistance,
providing a magnetic modulator (via a bridge circuit). The related Hall
effect, in which an e.m.f. is produced across a strip of material in a magnetic field, can be used either for field measurement (electric–magnetic–
electric modulator) or for current measurement (electric–electric modifier,
since magnetic energy is not consumed).

The effect can thus be used to provide transducers of all three basic
types.

10.4 Solution to exercises on chapter 4

4.3.1. An equivalent circuit is shown in figure 10.6.

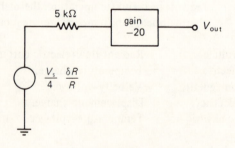

Figure 10.6

$$V_{out} = -20 \times \frac{4}{4} \times \frac{0.5}{10} \text{ per } °C$$

$$r = 1 \text{ V}/°C. \quad \text{(at operating temperature)}$$

4.3.2. Figure 10.7 shows the equivalent circuit. Note that a $1:1+1$ transformer was used (this is much easier to wind than the $1:\frac{1}{2}+\frac{1}{2}$ type used in deducing the general formulae).

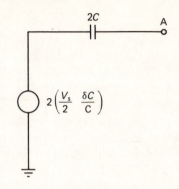

Figure 10.7

Charge amplifier: Gain (with respect to generator) is -4

$$\text{responsivity} = V_s \frac{\delta C}{C} \times 4 = 0.4 \text{ V/mm}$$

Non-inverting amplifier: Gain is $+100$ (strictly $10^6/(10^6 + 10^4)$)

$$\text{responsivity} = V_s \frac{\delta C}{C} \times 100 = 10 \text{ V/m}$$

However, this requires that the amplifier has a very high input resistance, substantially greater than the impedance of the capacitors (≈ 100 kΩ at 100 kHz).

10.5 Solutions to exercises on chapter 5

5.5.1. The main advantages of LVDTs are small size, ruggedness and tolerance of environmental conditions; the main disadvantages are relatively low responsivity, phase shifts (depending on frequency) and relatively high quadrature signals. For displacements in the μm region a device of small range (say, 1 mm) would be preferable, with responsivity, say, 10^3 or 10^4 V/m.

Variable area capacitance transducers have fairly low responsivity of about 100 V/m, but have very low quadrature signals, permitting high gain in the following amplifier. They tend to be relatively large and can be seriously affected by moisture and other environmental factors. They are very linear, apart from edge effects near the limit of their range. They usually have a range of ±2 or 3 cm, since the capacitances would be unduly low with smaller plates.

5.5.2. (a)

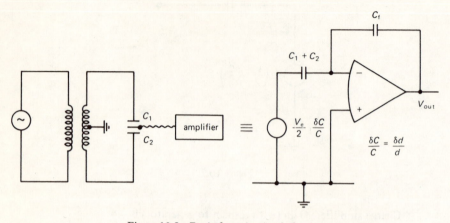

Figure 10.8 *Equivalent circuit of tiltmeter*

(b) $C_1 + C_2 = 2\dfrac{\epsilon A}{d} = 2 \times \dfrac{10^{-9}}{36\pi} \times \dfrac{10^{-3}}{5 \times 10^{-4}} = 35\ \text{pF}$

responsivity $= \dfrac{V_e}{2d} \times \dfrac{(C_1 + C_2)}{C_f} = \dfrac{2}{2 \times 5 \times 10^{-4}} \times \dfrac{35}{10} = 7000\ \text{V/m}$

A tilt of θ radians produces a displacement $L\theta$, so the responsivity in terms of tilt is 350 V/radian.

(c) The linearity will be poor unless the displacement of the bob is much less than the plate separation. A variable-area transducer would have much better linearity (though lower responsivity). Usually, however, such systems are operated in force-feedback arrangements, and the transducer is then used only as a null detector, so the linearity is much improved.

5.5.3. A resolution of 0.01 mm requires 1000 lines/cm which is rather high, and it would be easier to use gratings of 100 lines/cm and × 10 interpolation. An LED could be used as the source, with 2 or 4 photovoltaic cells attached to the 2-phase (or 4-phase) index. The separation between the gratings would have to be small (< 1 mm), so a good mechanical design is required.

Moiré gratings are the most obvious solution here, and are well suited to the application, since their main disadvantage (lack of absolute datum) is unimportant because the device would always be zeroed before each measurement.

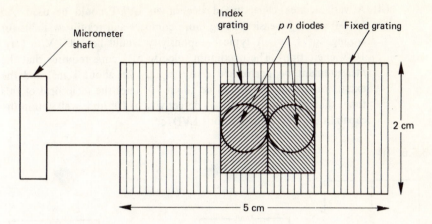

Figure 10.9 *Micrometer system*

5.5.4. The wind direction could be measured in many ways, using analogue transducers (resistive, capacitive or inductive angular types, synchros etc. followed by an A/D) or by digital devices (incremental angular encoders, radial Moiré gratings or absolute encoders). An accuracy of say ±5° would probably suffice, so a 5 or 6 track absolute encoder would be suitable and would avoid the need for a sense of direction indication (the vane could be allowed to rotate through many revolutions without ambiguity).

The wind speed could similarly be measured by analogue methods (tachometer and A/D) or digital methods (incremental encoders, toothed wheels etc.). A simple inductive proximity device feeding a frequency counter would probably be easiest, though an incremental encoder may be better for very low speeds, providing, say, 48 pulses per revolution (no indication of direction of rotation is needed). A (low) speed of 0.5 rev/s would still give an accuracy of better than 5 per cent with direct frequency counting.

5.5.5. The terms *responsivity*, *detectivity*, *range* and *accuracy* have been discussed in the text. Accuracy (sometimes called 'precision') means the absolute accuracy (that is, the difference between 'true' value and measured value), whereas detectivity is a relative term.

(i) This can be done with a variable-separation capacitive transducer but is rather beyond the limits of the best LVDTs. The range is very small but it is impractical to use a plate separation of less than 0.1 mm. With an excitation of, say, 10 V this gives a responsivity of 10^5 V/m, so a pre-amplifier of gain 100 is required. We have not dealt with noise in this book, but capacitive devices are essentially noiseless and the detectivity is set by the noise in the amplifier and can fairly easily be made equivalent to better than 10^{-12}/m (per Hz).

(ii) A variable-area capacitive device or an LVDT could be used. An LVDT would be easier, being more compact and readily available for a range of 1 cm. A typical responsivity would be 1000 V/m (say, 10 V excitation). The maximum velocity of 1 cm/s requires that the measurement be performed fairly rapidly (in about 1 ms) and the excitation frequency should be about 10 times the reciprocal of this time, that is, about 10 kHz. A detectivity of 10^5/m is well within the capabilities of even the author's LVDTs.

5.5.6. (i)

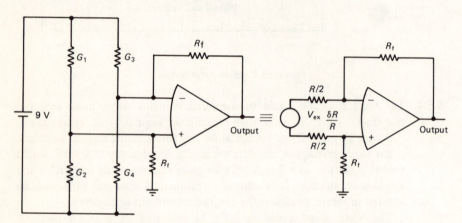

(gauge resistance = R)

Figure 10.10 *Equivalent circuit of strain gauge system*

$$V_{out} = V_{ex} \times \frac{\delta R}{R} \times \text{gain}$$

$$\text{GF} = \frac{\delta R/R}{\text{strain}} \text{ so } V_{out} = V_{ex} \times (\text{GF}) \times \text{gain (per unit strain)}$$

$$= 9 \times 2 \times 200$$
$$= 3600 \text{ V/unit strain}$$

(ii) Surface stress $= \dfrac{6fx}{bd^2} = \dfrac{6f \times 4 \times 10^{-2}}{10^{-2} \times 4 \times 10^{-6}} = 6 \times 10^6 \text{ N/m}^2 \text{ per N}$

Surface strain $= \dfrac{\text{stress}}{E} = \dfrac{6 \times 10^6}{2 \times 10^{11}} = 3 \times 10^{-5} \text{ per N}$

Amplifier output $= 3600 \times 3 \times 10^{-5} = 0.108 \text{ V/N}$
(= overall responsivity)

$$\text{Max. force} = \frac{9}{0.108} = 83.5 \text{ N} \quad \left(\text{corresponding strain} \approx \frac{1}{400}\right)$$

10.6 Solutions to exercises on chapter 6

6.3.1. The temperature of the handle will be in the range 20–100°C, mostly about 30°C, and an accuracy of about 1°C would be sufficient.

(i) Thermistors are suitable for the temperature range, though linearisation (or a look-up table) will be necessary if the range is more than about 10°C. They have a short time constant and could easily be fitted to the handle. A simple d.c. bridge would be satisfactory, including a linearising resistor.

(ii) Resistance thermometers could be used, the small Pt-film types being best. However, they are a little bulky for this application, though they could easily be fitted to the handle. Their resistance is low, so some compensating leads may be necessary.

(iii) Thermocouples are usually used for higher temperatures but could certainly be used here. Most types would be satisfactory, especially Chromel/Alumel, and most commercial units would give the required accuracy. They are a little bulky and not very well suited here.

All three transducers could be used, but thermistors are preferred.

6.3.2. 500–1000°C: Thermocouples or resistance thermometers should be used. Chromel/Alumel thermocouples feeding a commercial cold-junction compensated electronics unit would be easiest.

50–100°C: A thermocouple could be used here too, though the accuracy is near the limit of most units. Thermistors would be suitable with some linearisation or alternatively p-n diodes (calibration needed) or 'current sources'.

25–35°C: A thermistor would be satisfactory here (alternatively a p-n diode or 'current source' device). An accuracy of ±0.1°C is well within the capabilities of thermistors, but much more difficult with thermocouples and also with p-n diodes (since calibration is needed). A simple d.c. bridge would suffice.

6.3.3.

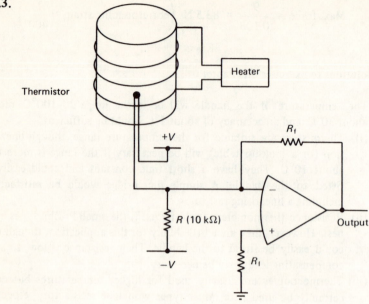

Figure 10.11 *Thermal system*

If the excitation is V_s the power dissipated is $V_s^2/4R$ which must be less than 10^{-4} W so $V_s \approx 2$ V. A simple d.c. thermistor bridge will be satisfactory since a detection level of about 0.01°C is required and most good operational amplifiers (for example, 741S) will not produce excessive drift (the resistor R_1 should be equal to that seen at the inverting input, which is $10\,k \parallel R_f$).

The output is given by

$$\frac{V_s}{4} \times \frac{\delta R}{R} \times \frac{R_f}{R/2} = \tfrac{1}{2} \times \frac{400}{10^4} \times \frac{R_f}{5 \times 10^3} = 1 \text{ V/°C}$$

so R_f should be 250 kΩ.

10.7 Solutions to exercises on chapter 7

7.4.1. A linear device is preferable, covering the visible region and extending a little into the infra-red, and with a fairly short time constant.
  (i) CdS. This is widely used in exposure meters because of its high responsivity. However, it is very non-linear, variable from device to device, slow, subject to drift and has a limited spectral response. In other words, it is not a good choice.

(ii) Silicon solar cell: Linear (in current), good spectral response (up to 1 μm) and very fast (much faster than required). Filters will be necessary to make measurements at different wavelengths, so some calibration will be needed (for each filter, and because the responsivity varies with wavelength in any case). A simple current amplifier is required. This is the best choice.

(iii) Pyro-electric cell. Responds well into the infra-red, but does *not* respond at d.c. *Not* suitable.

7.4.2. (i) Almost any type could be used, but a cheap photoconductive type such as CdS is most appropriate. A very coarse switching action is all that is needed and the exact level is not very critical.

(ii) A good infra-red response is essential since the peak emission will be at about 10 μm (see end of solution to exercise 7.4.4). Possible devices are PbS (response only to 3 μm), InSb (spectral response to 6 μm, but may need to be cooled), or a thermal type such as a thermistor bolometer or a pyro-electric cell (though does not respond at d.c.). The latter is probably the best choice here.

(iii) Most meters use CdS, since its responsivity is large and its spectral response similar to that of the eye, but it is non-linear and subject to drift. A better choice would be a silicon solar cell with a suitable filter ('eye-response' silicon cells are now available and could be used).

(iv) A fast response is required here and a silicon solar cell is the obvious choice, being linear in current responsivity and suitable in spectral response. A simple current amplifier can be used.

(v) A high detectivity is needed and a device suited to low light levels would be best, such as a photomultiplier or a PIN photodiode (or avalanche photodiode). PbS would be suitable if a response into the infra-red was needed, but a PIN diode would probably be the best choice.

7.4.3. Diameter of beam at detector $\approx 2 \times \dfrac{20}{60}$ so area $\approx 3 \times \dfrac{4}{9} \times \dfrac{1}{4} \approx \dfrac{1}{3}$ m^2

Radiation received $\approx 10 \times \dfrac{10^{-4}}{0.33} \approx 3 \times 10^{-3}$ μW

Signal current $= rW = 3 \times 10^{-10}$ A

Amplifier output $= 3 \times 10^{-10} \times 10^5 = 3 \times 10^{-5}$ V

This is a small signal but is fairly easily detected in a bandwidth of, say, 100 Hz, and a signal-to-noise ratio of more than 10 would be expected in practice.

7.4.4.

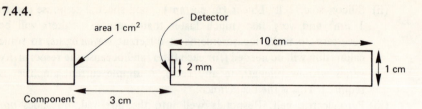

Figure 10.12 *Radiant system*

Radiation received. $W = \dfrac{\epsilon \theta a_1 a_2 T^4}{\pi d^2}$ where ϵ = emissivity (say, 0.5)

(assume $T \approx 300K$) $= \frac{1}{2} \times \dfrac{6 \times 10^{-8} \times 10^{-4} \times \pi \times 4 \times 10^{-6}}{\pi \times 9 \times 10^{-4} \times 4} \times 81 \times 10^8$

$$\approx 27 \ \mu W$$

Change per $^\circ$C $= \delta W = \dfrac{4W}{T} \times \delta T = 36 \times 10^{-8} \ W/K$

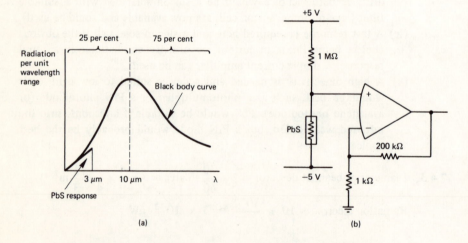

Figure 10.13 *(a) Black body curve. (b) Detector and amplifier*

Detector could be PbS (3 μm max.), InSb or pyro-electric (a.c. only). PbS would be most convenient, in a simple d.c. arrangement. The peak radiation will be at $2900/T$ μm ≈ 10 μm, so PbS will intercept only about 1/12 of the total radiation and its mean effective responsivity will be about 5 per cent of its peak value, say, about 500 V/W.

The output from the amplifier will be $36 \times 10^{-8} \times 500 \times 200 = 36 \times 10^{-3}$ V/°C. In practice a responsivity of, say, 0.1 V/°C would be appropriate, so a little more gain is needed. The system should be able to detect a change of about 1°C and this can be achieved fairly easily with PbS.

10.8 Solutions to exercises on chapter 8

8.5.1. The transfer functions are

(a) $\dfrac{x_r}{\ddot{x}_{in}}(s) = \dfrac{1}{s^2 + 2\xi\omega_0 s + \omega_0{}^2}$ and (b) $\dfrac{v_r}{v_{in}}(s) = \dfrac{s^2}{s^2 + 2\xi\omega_0 s + \omega_0{}^2}$

(i) Absolute velocity (10 Hz to 1 kHz)

Using (b) the response is flat (and of value unity) above the natural frequency f_0. We need to measure relative velocity v_r in a system with natural frequency less than 10 Hz (to avoid the peak produced at f_0). A simple cantilever beam can be used with a magnet/coil sensing system (better with a fixed magnet).

(ii) Absolute acceleration (10 Hz to 10 kHz)

Using (a) the response is flat (and of value $1/(\omega_0{}^2)$) below the natural frequency. We need to measure relative displacement x_r below f_0, so f_0 should be greater than 10 kHz. A piezo-accelerometer is necessary to achieve this high value of f_0, and does not operate down to d.c. when used as a displacement sensor. The design of the charge amplifier will determine the lower cut-off frequency.

(iii) Absolute displacement (0–100 Hz)

Equation (b) can be written in terms of x_r and x_{in} and permits absolute displacement to be measured (using a relative displacement transducer) *above* f_0. The only way to obtain absolute acceleration from d.c. upwards is to use relation (a) and double-integrate the acceleration. An open-loop system with $f_0 > 100$ Hz could be used, but drift problems are severe (because of the double integration) and closed-loop devices are preferable. A force-feedback accelerometer with closed-loop resonant frequency > 100 Hz (open-loop $f_0 \approx 10$ Hz) would be used.

8.5.2. Velocity ⌐ translational ⌐ relative: linear velocity transducers (LVTs) correlation methods

absolute: seismic mass (velocity meters and accelerometers)

rotational   ⌐ relative: tachos
timing methods (mechanical, optical)
correlation methods

absolute: seismic mass (velocity meters and accelerometers)

(i) This is a simple measurement of relative rotational velocity. A timing method (involving a proximity sensor, small magnet, toothed wheel etc.) is best. It will be necessary to use the counter in period mode (that is, gate the interval clock by the pulse produced once per rotation) to obtain the required accuracy, or to use a toothed wheel or equivalent with 10 teeth.

(ii) This is a straightforward measurement of relative translational velocity, and can be done with a simple magnet/coil system or preferably an LVT (because of the range required). Off-the-shelf LVTs are available for the specification required.

(iii) This requires an absolute translational device. Seismic-mass velocity meters having a natural frequency < 5 Hz are available and will provide a flat response from 5 Hz to well above 100 Hz. The maximum velocity is $10^{-4} \times 2\pi \times 10^2 \approx 0.1$ m/s, which is not excessive.

(iv) The requirement of a 'remote' measurement suggests that a correlation method should be used. Some commercial systems are available but a purpose-built system may be required, using optical reflection to produce the signals, and a binary correlation algorithm would be preferred. If the reflection points were 10 cm apart this would correspond to 10^{-2} s, so the data rate required is fairly high. A readout once every few seconds would be possible.

8.5.3.
(i) Compliance $= \dfrac{\text{deflection}}{\text{force}} = \dfrac{4l^3}{Ebd^3} = 2.5 \times 10^{-4}$ m/N

$$\omega_0{}^2 = \frac{1}{MC} = 4 \times 10^4, f_0 = 31.8 \text{ Hz}$$

(ii) The equivalent circuit is shown in figure 10.14.

$$V_0 = \frac{V_s}{2} \times \frac{\delta R}{R} \times \frac{R_f}{R/2} = \frac{V_s}{2} \times 2 \times \frac{R_f}{R/2} \times \text{strain} = 2000 \times \text{strain}$$

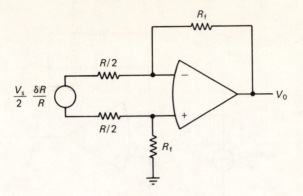

Figure 10.14

Below resonance $\dfrac{x_r}{x_{in}} = \dfrac{1}{\omega_0{}^2}$ so $x_r = \dfrac{1}{\omega_0{}^2}$ per ms^{-2}

Strain $= \dfrac{stress}{E} = \dfrac{6fx}{Ebd^2} = 12 \times 10^{-5} \; f\left(f = \dfrac{x_r}{C}\right)$

$\qquad = 0.48/m$

$V_{out} = \dfrac{2000 \times (0.48)}{4 \times 10^{-4}} = 0.024 \; V/ms^{-2} \; (= \text{responsivity})$

(iii) $\pm 10 \; V = 0.024 \times a_{max}$ so $a_{max} = 400 \; ms^{-2} \approx 40 \; g$

(the deflection produced ≈ 1 cm, which is rather high).

8.5.4. $E = \dfrac{stress}{strain} = \dfrac{f/A}{x/t}$ where $x = $ displacement and $t = $ thickness.

Compliance $C_m = \dfrac{x}{f} = \dfrac{t}{EA} = 2.5 \times 10^{-11} \; m/N$

$\omega_0{}^2 = \dfrac{1}{MC} = 8 \times 10^{11}$, so $f_0 \approx 142 \; kHz$

$C = 2\dfrac{\epsilon\epsilon_0 A}{t} \approx 16 \; pF$

$V_g = \dfrac{2K}{C} \; x$ where $K = \dfrac{dEA}{t} = \dfrac{d}{C_m}$ (see equation (5.2))

$V_g = \dfrac{2d}{CC_m} \; x = 5 \times 10^9 \; V/m$

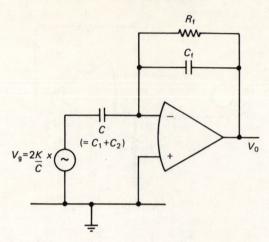

Figure 10.15 *Equivalent circuit of piezo-electric sensor*

$$V_0 = V_g \times \frac{sCR_f}{1 + sC_fR_f} \approx V_g \times \frac{C}{C_f} \text{ for } sC_fR_f \gg 1 \ (\approx 160 \text{ Hz})$$

Mid-frequency acceleration responsivity =

$$\frac{1}{\omega_0{}^2} x = \frac{1}{\omega_0{}^2} \times V_g \times \frac{C}{C_f} \approx 39 \text{ mV/ms}^{-2}$$

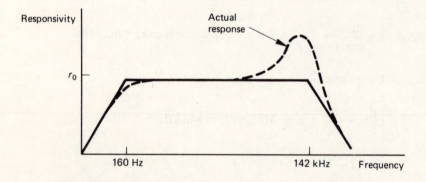

Figure 10.16 *Responsivity of piezo-accelerometer against frequency*

8.5.5. The compliance C_m =

$$\frac{4l^3}{Ebd^3} = \frac{4 \times 125 \times 10^{-6}}{2 \times 10^{11} \times 2 \times 10^{-2} \times 64 \times 10^{-9}} \approx 2 \times 10^{-6} \text{ m/N}$$

Deflection for force of $1000 \text{ N} \approx 2 \times 10^{-6} \times 10^3 \approx 2 \text{ mm}$

(i) Strain gauges. Max. surface stress =

$$\frac{6fx}{bd^2} = \frac{6 \times 10^3 \times 2 \times 10^{-3}}{2 \times 10^{-2} \times 16 \times 10^{-6}} \approx 0.4 \times 10^8 \text{ N/m}^2$$

Max. strain =

$$\frac{0.4 \times 10^8}{2 \times 10^{11}} \approx 2 \times 10^{-4}$$

Strain gauges are very suitable here. They have no particular disadvantages, apart from temperature effects which can be reduced by using several gauges as discussed in the text. Metallic foil types would be the best solution.

(ii) Piezo-crystals are *not* suitable here, as they do not respond to displacement at d.c., and a d.c. system is specifically stated.

(iii) Variable-separation capacitive transducers could certainly be used, though a rather wide plate spacing is needed ($> 2 \text{ mm}$). However, it would be much more complicated to attach them than strain gauges, and they are a very poor choice here.

8.5.6.

Beam: $\dfrac{\theta}{T} (s) = \dfrac{C\omega_0{}^2}{s^2 + 2\zeta\omega_0 s + \omega_0{}^2}$ (angle θ, torque T in rotational system)

$$\omega_0{}^2 = \frac{1}{JC} \text{ where } J = \frac{Ml^2}{3} = (0.1)\frac{(0.1)^2}{3} = \frac{1}{3} \times 10^{-3}$$

$$4\pi f_0{}^2 = \frac{1}{\frac{1}{3} \times 10^{-3} \times C} \text{ so } C = \frac{3}{4\pi^2 \times 25 \times 10^{-3}} \approx 3 \text{ radian/N m}$$

For translation, $x = l\theta$ and $f = \dfrac{T}{l}$ ($l = 10$ cm), therefore

$$\frac{x}{f} (s) = \frac{C\omega_0{}^2 l^2}{s^2 + 2\zeta\omega_0 s + \omega_0{}^2} \qquad \text{(deflection } x\text{, force } f \text{ in translational system)}$$

Displacement transducer: Variable-separation capacitive or LVDT.
     With 1 mm separation and 1 V excitation and $\times$ 10 pre-amp, $r \approx 10^4$ V/m.

Force-feedback: Voltage/force coefficient $= \dfrac{10}{R_f} \text{ N/V.}$

For high LG, overall responsivity $= \dfrac{R_f}{10}$ so $R_f = 100 \ \Omega$

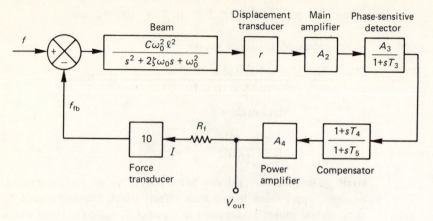

Figure 10.17 *Force-feedback system*

Loop-gain: $3 \times 10^{-2} \times 10^4 \times A_2 \times 1 \times 1 \times 1 \times \dfrac{1}{10}$ (taking A_3 and

A_4 unity)

$$\approx 30\, A_2 \text{ so } A_2 \approx 3$$

Compensation: Simple phase-lead. Integral control preferable.

Calibration: Add known mass on open loop and balance via force-transducer, detecting balance with displacement transducer.

8.5.7.   (i)   This can be done by a laser doppler system or by a correlation method – in either case it is going to be rather expensive. The laser system would be appropriate if there was a very wide flow range (up to high values) and if high accuracy is needed. Otherwise a correlation system, involving two separated beams of light passed through the tube, would be appropriate. The photodetector outputs can be digitised into binary values for easy interfacing and fast computation.

      (ii)   The rate of flow will vary with depth but it should be assumed that an average value is required. A 'weir' is probably the simplest method, measuring the height difference produced (rate $\propto$ (height)$^{3/2}$). If such an obstruction is not permissible a turbine could be used, adjusting the depth to give an average value. Alternatively, if no contact or disturbance is possible an optical correlation method could be used, though it would measure the velocity at the surface.

   (iii)   Small turbines in short tubes would probably be best here, since they could easily be placed at any depth and could also provide directional information by orienting the cylinder for maximum flow. A simple proximity detector providing a digital readout could be used. Other possible transducers are orifice plates (in cylinder), though some disturbance in flow will occur, or drag-force devices.

10.9 Solutions to exercises on chapter 9

9.5.1. (i) The main advantages of optical-fibre sensors are their immunity from electromagnetic interference and the ease with which they can be linked into an optical communication system. There are no particular disadvantages, though many of the sensing applications are somewhat artificial, the fibre being used essentially as a light guide. One interesting development is that the power for a sensing system can sometimes be fed down the fibres themselves.

The main properties of light suitable for such sensors are intensity, phase, polarisation, wavelength and spectral distribution, and the main modulating effects are piezo-absorption, stress birefringence, an electro-optical effect, a magneto-optical effect, the Faraday effect and some thermal effects.

The most widely used devices are intensity modulators, often employing the relative motion of two fibres, and sometimes using a 'microbending' arrangement (piezo-absorption) to measure force or other parameters derived from it. Phase modulators are not much used as sensors, though the fibre-optic gyroscope uses this effect. However, polarisation modulators, using Faraday rotation, are a valuable non-contacting means of measuring large currents.

(ii) Optical processing involves transforming and manipulating optical data by optical means, a particularly good example being afforded by an optical Fourier Transform whereby the light distributions in the front and back focal planes of a lens are a Fourier pair. Such systems were developed well before the recent microelectronics developments but suffered from difficulties in input and output of data. Indeed, the most successful applications (deblurring photographs and character recognition) did not use real-time data. To some extent the development of microprocessors has reduced the investigation of optical processing, since it is now relatively easy to digitise a picture and carry out a fast Fourier Transform. However, fundamental limits in size and speed of silicon-based devices are gradually being reached and optical systems, which have inherently very high data rates, may then receive a new impetus. 'Optical computers' are certainly a long way off, since many of the basic components simply do not exist, but there is sure to be a valuable 'spin-off' from developments in communication systems. The 'direct' coupling of fibres to conventional computers is relatively easy, of course, and more and more optical processing will be carried out before the data are transferred to the computer, which is a first step towards a complete optical processor.

9.5.2. The particular advantages of resonator sensor systems are that they produce a frequency output proportional to the quantity of interest, suitable for interfacing to computer systems, and that their stability is determined by mechanical components and is therefore high and little affected by electrical interference.

Several physical parameters can be measured by such systems. Devices may be classified into wires, beams and cylinders. Wire systems were the first to be used, usually for measurement of force or pressure. Vibrating beams may similarly be used for force measurement, but can conveniently be used for liquid density, level or viscosity by placing a part of the beam in the liquid of interest. It is also possible to measure flow by using such a beam in a tube. Very small beams made from quartz or etched from silicon can be produced, often in a double tuning-fork arrangement, and are used in miniature pressure or vibration sensors. Vibrating cylinders are similar to vibrating beams, but are well suited to density and flow measurement by designing a suitably vibrating section of tube.

A rather different principle is used for the detection of some chemical quantities, a vibrating quartz crystal having a suitable absorbing film deposited on one face, and the change in frequency monitored.

It is often suggested that a resonator sensor system is inherently superior to a conventional analogue system, because it is 'inherently digital'. In fact it is nothing of the sort, but is simply of a form that is easy to digitise accurately. However, there are severe costs to be paid for this, in the form of inherent non-linearity and severe temperature dependence, and in many cases one would be better off with a conventional analogue system and a simple A/D unit. The most promising devices are the miniature beams and quartz crystals with absorbing layers.

9.5.3. (i) The properties of silicon that make it suitable for VLSI are that it can be produced in very pure large crystals and that its oxide provides an excellent mask for selective doping and etching. These properties make it excellent both for the production of large numbers of very small circuit elements and for the physical construction of small structures for use as resonator (or other) sensors. In addition, there are a number of useful effects which can be used for self-generators or modulators.

The main self-generator effects are the Seebeck, photovoltaic and galvano-electric. It lacks a suitable mechanical or magnetic effect, but because of the ease of masking and doping a layer of suitable material can easily be added. There are modulating effects for all the main energy forms, the most important being piezoresistive, resistance/

temperature, photoconductive and magnetoresistive/Hall. The use of silicon-based solid-state sensors will certainly increase substantially in the next few years.

(ii) The premises on which the theory is based appear very reasonable. Carbon forms an enormous number of compounds and appears to be the only element capable of forming the complicated molecules necessary for the natural development of organic life. Although silicon-based compounds could not develop naturally in this way, silicon is a remarkably good material for producing miniature electrical circuits, as discussed above. It seems likely that as packing densities are increased units will be produced that considerably surpass the storage of the human brain, and with an access time many orders of magnitude shorter. Of course, what constitutes intelligence and whether it can ever be truly artificial is another matter, and not within the intended scope of this book. However, to be on the safe side, the writer's new year resolution is to be kinder to his home computer!

References and Bibliography

It would not be very difficult to produce an enormous list of references, because almost every physical effect or transducer mentioned could be separately referenced. However, it was thought to be more useful to give specific references to only some of the more recent developments, but to give a general bibliography to the subject as a whole. Most of the books mentioned are of course books on measurement, since there are very few available that concentrate on sensors and transducers.

References

Asher, R.C. (1983). 'Ultrasonic sensors in the chemical and process industries', *J. Phys. E*, Vol. 16, No. 10, pp. 958-64.

Beck, M.S. (1983). 'Cross correlation flowmeters', *Instrument Science and Technology*, Vol. 2 (ed. B. Jones), Adam Hilger, Bristol, pp. 89-106.

Brignell, J.E. and Dorey, A-P. (1983). 'Sensors for microprocessor-based applications', *J. Phys. E*, Vol. 16, No. 10, pp. 952-8.

Culshaw, B. (1982). 'Optical fibre transducers', *Radio and Electronic Engineer*, Vol. 52, No. 6, pp. 283-91.

Doebelin, E.O. (1966). *Measurement Systems: Application and Design*, McGraw-Hill, Maidenhead, Berks.

Harmer, A.L. (1982). 'Principles of optical fibre sensors and instrumentation', *Measurement & Control*, Vol. 15, April, pp. 143-51.

Jones, B.E. (1977). *Instrumentation, Measurement and Feedback*, McGraw-Hill, Maidenhead, Berks.

Middelhoek, S. and Noorlag, D.J.W. (1981a). 'Silicon microtransducers' *J. Phys. E*, Vol. 14, 1343–52.

Middelhoek, S. and Noorlag, D.J.W. (1981b). 'Three-dimensional representation of input and output transducers', *Sensors & Actuators*, Vol. 2, 29–41.

Shearer, J.L., Murphy, A.T. and Richardson, H.H. (1971). *Introduction to System Dynamics*, Addison-Wesley, London

Van Dijck, J.G.R. (1964). *The Physical Basis of Electronics*, Centrex, Eindhoven/ Macmillan, London, pp. 31–44.

Bibliography

Chesmond, C.J., *Control System Technology*, Edward Arnold, London 1982.
Mostly concerned with control technology but a useful section on transducers.

Dance, J.B., *Photoelectronic Devices*, Iliffe, Guildford, 1969.
Useful treatment of semiconductor fundamentals and optical devices.

Doebelin, E.O., *Measurement Systems: Application and Design*, McGraw-Hill, Maidenhead, Berks., 1966.
Very comprehensive coverage of measurement systems (700 pages), excellent as reference book.

Giles, A.F., *Electronic Sensing Devices*, William Clowes, Colchester, 1966.
Provides a useful physical background and includes chemical sensors.

Gregory, B.A., *An Introduction to Electrical Instrumentation and Measurement Systems*, Macmillan, London, 1973.
Concentrates on electrical instrumentation but much useful material.

Handbook of Measurement Science, Vols 1 and 2 (ed. P.H. Sydenham), Wiley Chichester, 1982/3.
Very thorough review of theoretical fundamentals (vol. 1) and practical applications (vol. 2), about 2000 pages in all.

Instrument Science and Technology, Vols 1 and 2 (ed. B.E. Jones), Adam Hilger, Bristol, 1982/3.
Collected review papers from *J. Phys. E* in the instrument science field. There is

also a very useful special issue of *J. Phys. E* (Vol. 16, No. 10, 1983) containing the invited papers from the conference on 'Sensors and their Applications' at UMIST in 1983.

Jones, B.E., *Instrumentation, Measurement and Feedback*, McGraw-Hill, Maidenhead, Berks., 1977.
Integrated and up-to-date account of the measurement field (283 pages).

Neubert, H.K.P., *Instrument Transducers*, OUP, London, 1963.
Detailed design criteria of various transducers, mostly inductive.

Oliver, B.M. and Cage, J.M., *Electronic Measurements and Instrumentation*, McGraw-Hill, New York, 1971.
Useful reference book.

Woolvet, G.A., *Transducers in Digital Systems*, Peter Peregrinus, Hitchin, Herts., 1977.
Useful survey of digital devices, including encoders, resonator systems and ADC/DACs.

Index